Hydrogen Projects

Legal and Regulatory Challenges and Opportunities

Author
Dalia Majumder-Russell

Managing director
Sian O'Neill

Hydrogen Projects: Legal and Regulatory Challenges and Opportunities
is published by

Globe Law and Business Ltd
3 Mylor Close
Horsell
Woking
Surrey GU21 4DD
United Kingdom
Tel: +44 20 3745 4770
www.globelawandbusiness.com

Printed and bound by Ashford Colour Press Ltd

Hydrogen Projects: Legal and Regulatory Challenges and Opportunities

ISBN 9781787424425
EPUB ISBN 9781787424432
Adobe PDF ISBN 9781787424449

DISCLAIMER
This publication is intended as a general guide only. The information and opinions which it contains
are not intended to be a comprehensive study, or to provide legal or financial advice, and should
not be treated as a substitute for legal advice concerning particular situations. Legal advice should
always be sought before taking any action based on the information provided. The publishers bear
no responsibility for any errors or omissions contained herein.

Table of contents

I. Introduction

1. The background

Climate change commitments made by governments and corporates under the 2015 Paris Agreement are driving home the message that cutting greenhouse gas (GHG) emissions only through carbon offsets and renewable electricity will not be enough. Can the world's lightest, most abundant element save the day? Many are hailing hydrogen as the revolutionary, clean energy vector that can help to:

- decarbonise industrial processes;
- heat and cool buildings;
- refuel transport where battery technologies cannot; and
- manage intermittent electricity grids.

The use of hydrogen is not new. It has been commonly used in industrial processes since the Swiss inventor, Isaac de Rivaz, created the first internal combustion engine using hydrogen in 1806. Since then, hydrogen has been a signature feature of space programmes and industrial processes ranging from petrochemistry to metallurgy to pharmaceutical. Yet its application in transport, and efforts to make it low carbon, are new. The drivers for this are twofold:

- national and international commitments to decarbonise, combined with the falling costs of renewable electricity generation; and

- 'green' hydrogen-producing electrolysers which offer a viable alternative to traditional hydrogen production.

Figure 1 provides an overview of the means commonly used to produce 'green' and 'blue' hydrogen.

Alongside our understanding of the different means of producing and using low-carbon hydrogen, the so-called 'hydrogen rainbow' has evolved, with different colours assigned to hydrogen depending on its method of production – see Table 1, in chapter III, for an explanation of the various colours. To be clear, despite references in common parlance to the different colours of the hydrogen rainbow, hydrogen in its elemental form has no colour or smell. Blue and green are the most common low-carbon hydrogen 'colours', and these will form the focus of this Special Report.

Currently, hardly any hydrogen produced is low carbon. Instead, so-called 'grey' hydrogen, which emits carbon dioxide (CO_2) into the atmosphere at the rate of 9.3kg of CO_2 produced per kilogram of hydrogen produced, is the most commonly consumed type of hydrogen. This, according to the International Energy Agency (IEA), results in emissions of around 830 million tonnes of CO_2 per year,

Figure 1. Production and uses of 'green' and 'blue' hydrogen[1]

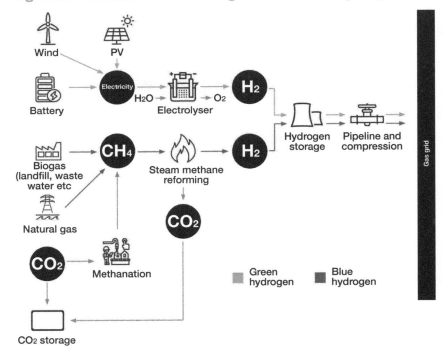

Source: Adapted from Fig 1 in Quarton and Samsatli, "Power-to-gas for injection into the gas grid" (see note 1).

"The development of low-carbon hydrogen therefore requires targeted subsidy arrangements, supportive procurement programmes and a stable and well-defined regulatory market within which the developers, investors and financiers of such projects can operate."

equivalent to the CO2 emissions of the United Kingdom and Indonesia combined.[2] Clearly, much can be achieved from an environmental perspective by switching current hydrogen production methods to low-carbon options of blue or green hydrogen.

The role that low-carbon hydrogen can play in the global energy mix will depend on stakeholder ambitions and its potential applications in the energy, heating, transport and industrial sectors. As we examine in this Special Report, the role of stakeholders such as central and local governments is key in stimulating the so-called 'hydrogen economy'. Left to market forces alone, new technologies such as low-carbon hydrogen projects, with their high upfront costs and lack of cost competitiveness, would struggle against established ways of making hydrogen. The additional costs of carbon capture, usage and storage technology (CCUS) needed to capture CO2 emissions constitute an upfront cost not present in grey hydrogen production. The development of low-carbon hydrogen therefore requires targeted subsidy arrangements, supportive procurement programmes and a stable and well-defined regulatory market within which the developers, investors and financiers of such projects can operate.

This Special Report also considers which sectors would benefit from using or switching to low-carbon hydrogen. We consider the challenges of using it, along with the opportunities that arise in the transport and industrial manufacturing sectors, and the role it can play in heating/cooling and energy system management applications.

Finally, because no discussion about low-carbon hydrogen is complete without analysing the issues for producers and users in the context of equity and debt investment, we focus on common themes seen in the development of recent low-carbon hydrogen projects.

This Special Report naturally has a European and UK focus, given that these geographies are among the most advanced in terms of blue and green hydrogen projects. In 2018 the UK's Climate Change Committee (CCC) identified low-carbon hydrogen's role in achieving the UK's decarbonisation targets.[3] The UK government's December 2020 energy white paper followed, amplifying the role that the government sees hydrogen playing in its energy mix.[4] As part of this, the government's R&D body, UK Research and Innovation (UKRI) awarded £171 million of funding across nine hydrogen projects in March 2021[5] and (as at 1 July 2021) the UK government is working on developing business models that will support investment in CCUS and hydrogen projects.[6]

Similarly, since the European Commission's July 2020 announcements on energy systems integration and hydrogen strategy for a climate-neutral Europe, the landscape for projected development and investment in low-carbon hydrogen projects in the European Union has never looked more promising.[7, 8]

The first of the EU's ambitious goals is to be able to produce one million tonnes of renewable hydrogen by 2024, increasing to 10 million by 2030.[9] The EU's network operators are already discussing blending methane and hydrogen gas for shipping across national borders, and some of the key advances in low-carbon hydrogen development, such as the operation of the world's largest (10MW) green hydrogen production facility (started in Germany in 2021)[10] and the world's first hydrogen-powered trains, made by Alstom (operating in Germany since 2016), have taken place in European Union member states.

However, this energy transition is global and, like other energy transitions before it, will involve a global response with the emergence of new players and the repositioning of current centres of power and dominance. Already, according to the Hydrogen Council, there are over 30 countries with hydrogen roadmaps, and 228 large-scale hydrogen projects announced across the value chain, with 85% located in Europe, Asia and Australia.[11] Some of the most advanced hydrogen projects are currently being developed in East Asia and the Asia-Pacific regions. This is only likely to accelerate, with countries like

Figure 2. Trends in hydrogen use, 1980–2018[12]

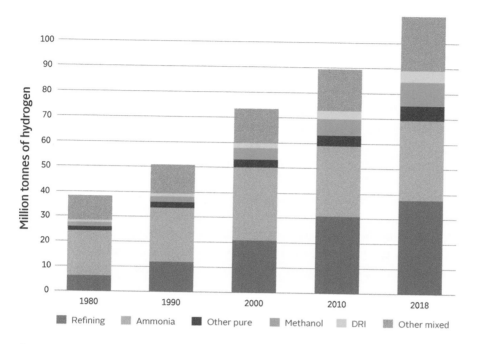

Source: Reproduced by permission of the International Renewable Energy Agency (IRENA) (see note 12).

Japan, China and South Korea all making pledges to become carbon-neutral economies over the coming decades.

Like the other energy transitions, a hydrogen-based energy transition will not happen over a short period of time and will require sustained efforts on a number of fronts. Hydrogen projects will need to compete against other new energy transition technologies, including electrification, and be applied in sectors and geographies that are best suited to their use as part of a complementary solution to changing global energy needs. Therefore, whether the world's lightest and most abundant element will become a key part of the next energy mix will depend on what is done to overcome the challenges – be they legal, economic or technical – facing hydrogen projects being developed today, so as to capitalise on the tantalising potential of this resource.

2. What do we mean when we talk about hydrogen?

Hydrogen has come to the fore in discussions around the global energy transition, and efforts to address the adverse impacts of climate change, due to its molecular properties and the ability of the technology to interact with other global efforts in the decarbonisation arena. Hydrogen has the highest energy content by weight of any common fuel and is a high-efficiency, low-polluting fuel.

Many predict that hydrogen will form a key part of the global energy mix. It may be used to support a number of current energy consumption markets including fleet or long-haul transport, chemical production processes and manufacturing of materials such as iron and steel, as well as to help achieve increased stability in an electricity system powered by intermittent generation technologies. Additionally, such applications can help to address air quality and energy security issues through reduction of GHGs in the atmosphere. And finally, producing hydrogen using renewable electricity generation can constitute a near-zero-carbon production method, thus decarbonising other processes where it can be applied.

There are several reasons why the energy sector is getting excited about hydrogen as an alternative energy source. While hydrogen has traditionally been used as a feedstock for industrial processes such as ammonia synthesis and the refining of crude oil, development over the last few years has shown a number of other useful applications. This is stimulated in part by a move towards electricity generation from intermittent renewable sources, the value of which can be enhanced by using the surplus energy to create hydrogen, which can in turn be used at times of higher energy demand; and in part by the application of hydrogen as a fuel across the transportation sector. Improvements in hydrogen technologies, and increasing government support for low-

"While hydrogen has traditionally been used as a feedstock for industrial processes such as ammonia synthesis and the refining of crude oil, development over the last few years has shown a number of other useful applications."

carbon options to diversify and decarbonise the energy industry, mean that hydrogen is receiving serious consideration as an alternative energy source with the potential to contribute to net zero targets. In the United Kingdom the government (in its Clean Growth Strategy)[13] and the UK Committee on Climate Change[14] have identified hydrogen as the most cost-effective option for decarbonising parts of the UK energy system.

Two key developments have contributed to the growth of hydrogen in recent years:

- The cost of hydrogen from renewables has decreased, and continues to fall.
- The urgency of GHG emission mitigation has increased.

As such, many countries are taking action to decarbonise their economies. The hydrogen debate has been around for decades, but recently the focus has shifted towards its application.

The International Renewable Energy Agency (IRENA) reports that around 120 million tonnes of hydrogen are produced annually, which represents about 4% of global final energy and non-energy use.[15] Most of the hydrogen produced is transported (either by pipeline systems or by road) for use in industrial processes such as the production of ammonia as a nitrogen fertiliser, for chemical production processes, and in oil refining as part of transport fuel production.

This means that although much discussion has centred around the production, transportation and use of low-carbon hydrogen, there is very little hydrogen currently in use for such purposes. Instead, as noted above and described further in chapter III, most of the hydrogen produced and consumed is known as 'grey' hydrogen.

Despite the relatively modest usage of low-carbon hydrogen today, when most developers, operators, investors and financiers talk about hydrogen projects in the global energy context, and especially in the context of decarbonisation, it is low-carbon hydrogen that is being discussed.[16] Of these the most common are blue and green. (See Table 1 in chapter III for further information about the different ways of producing hydrogen.) As new applications come to the fore and the barriers currently faced by such projects are overcome, it is these projects that will have an important role to play in the energy transition. Amongst its potential uses, low-carbon hydrogen may be used as an alternative to methane gas, for fuelling a range of energy needs, and as a feedstock in industrial and chemical processes, as well as in domestic contexts such as heating/cooling and transport.

Because hydrogen is a versatile energy vector, unlike electricity it can be stored, transported and converted for use in a range of specific applications, satisfying their particular requirements. In addition, hydrogen can avoid the CO_2 emissions that accompany unabated fossil fuel use. Furthermore, hydrogen produced using biomass with CCUS can, in some circumstances, offer negative GHG emissions where the biomass has acted as a CO_2 sink during its lifetime and the GHGs emitted during hydrogen production are captured and stored.

3. Challenges and opportunities

In this Special Report, we will examine the legal and regulatory challenges and opportunities of developing hydrogen projects within the context of how hydrogen is anticipated to be used in the coming decades. While no one can predict the future with full certainty, the developments and innovations of today give a good indication of the major infrastructure, applications and uses that are aspired to in the future. At present, three major areas of end use are suitable for transition to hydrogen:

- transport;
- industrial processes; and
- heating and cooling.

3.1 Transport

While battery-powered vehicles (primarily using lithium-ion batteries) are seen as a viable technology for passenger vehicles, delivering a low-carbon solution for larger road, marine and rail vehicles is more challenging. Hydrogen has been proposed in the industry as a potential solution, as hydrogen fuel cells are well suited to heavy-duty trucks, fleet vehicles and buses as well as being able to power ships, rail transport and, in time, aircraft.

The only emission from a hydrogen fuel cell-powered vehicle (or HF(E)V) is water vapour, which means the technology can be used to improve air quality in cities, where public resistance to pollution from vehicle (and industrial) emissions is rising. In terms of other emissions, it depends on whether the hydrogen produced is low carbon (see Figure 1 above) or generated using unabated fossil fuels.

While fuel cell production costs remain high, hydrogen fuel cell vehicles have some key performance advantages over battery electric vehicles. The range of most hydrogen fuel cell vehicles is around three times that of the average battery electric vehicle, and refuelling times are significantly shorter. For example, Hyundai's hydrogen fuel cell-powered Nexo has a 414-mile (666-km) range compared to, say, the 150-mile (240-km) range of the Nissan Leaf Acenta li-ion vehicle. In addition, like most hydrogen fuel vehicles, the Nexo can be filled almost as quickly as a petrol or diesel car.[17]

3.2 Industrial processes

Hydrogen is already used to a significant extent as a feedstock in industrial and manufacturing processes. Around 90% of the hydrogen market is accounted for by these applications. As mentioned above, it is used in chemical engineering, refining and metal processing among other industrial uses. It is predicted that this sector will continue to grow year on year to service the demands of a growing global economy, and therefore decarbonisation of industrial processes is key to achieving national and international climate change targets. Accordingly (and as discussed further in chapter V), replacing grey hydrogen with green or blue hydrogen in industrial and manufacturing processes is key to decarbonising these areas, and a number of companies are working on technologies to incorporate greater use of green or blue hydrogen into their processes.

Case study: hydrogen for iron ore reduction[18]

Accounting for 7–9% of global CO_2 emissions and in increasing demand, the steel production industry is turning to innovative solutions to accelerate emissions reductions. The use of hydrogen in steel production facilities is being explored as an increasingly viable way to support decarbonisation of this heavily emitting sector. In Austria, partners Mitsubishi Heavy Industries and Voestalpine are building the world's largest green hydrogen steel plant to pilot hydrogen-fuelled iron ore reduction processes. The reduction process is an essential but carbon-intensive step in steel production that accounts for a large proportion of the industry's emissions. The Austrian plant will use hydrogen as an alternative to coal to facilitate the removal of oxygen from the iron ore in a direct reduction process. The anticipated production capacity of the project is 250,000 tonnes of steel product per year, and it has the potential to attain net zero CO_2 emissions. To further increase the project's energy efficiency, a heat recovery system will be implemented.

3.3 Heating and cooling

Keeping buildings cool in the warmer months and warm in the cooler ones is a perennial challenge worldwide. Currently, air conditioners are mostly powered by electricity whereas heating is often achieved using methane gas. Heating and cooling consume half of the EU's energy generation.[19] If Europe wants to decarbonise heating and cooling it must consider either:

- the electrification of heating using heat pumps or electric heaters; or
- the introduction of renewable gases such as hydrogen.

A number of countries are exploring the latter option either by blending hydrogen with methane in existing gas networks, or adapting gas networks for the transportation and storage of pure hydrogen. By this 'greening of the gas grid', gas networks are giving serious consideration to their ability to deliver hydrogen instead of methane gas to homes and businesses. For example, a number of projects and trials are being undertaken to blend hydrogen into the existing (methane) gas supply which don't require changes to gas appliances or pipework, while still cutting carbon emissions. The United Kingdom is trialling up to 20% hydrogen in the gas grid without any infrastructure modifications. Some other countries have grids that would be compatible with 100% hydrogen.

4. Hydrogen and net zero

Hydrogen will be part of efforts to mitigate emissions in the coming decades. IRENA's Renewable Energy Roadmap (REmap) analysis indicates a 6% hydrogen share of total final energy consumption by 2050[20] and the Hydrogen Council's roadmap suggests that an 18% share can be achieved in the same time frame.[21] While it may not be clear exactly how much hydrogen will contribute to the overall energy share by 2050, it's clear that it is key to mitigating emissions and achieving net zero goals.

Today, almost all hydrogen is produced from natural gas and coal; that's to say, it is grey or brown/black hydrogen. This means that swift and deep changes are needed to produce more hydrogen from low-carbon sources if the decarbonisation aims are to be realised. Production of hydrogen from low-carbon energy is currently costly, but these costs are falling rapidly. Additionally, as hydrogen is mostly used in industrial processes, a shift to using it in sectors such as transport, heating and power system management is needed.

The development of a hydrogen economy also depends on the development of hydrogen infrastructure – new and upgraded pipelines and efficient and economic shipping solutions. Government support, too, is needed; indeed, in many countries regulations currently limit the development of a hydrogen economy.

II. Stakeholders

1. Introduction

This chapter provides an overview of the stakeholders involved in the development and use of low-carbon hydrogen projects. While each country will have its own legal and regulatory framework which influences how the energy transition is progressing and the role of low-carbon hydrogen within it, the interests of stakeholders have some commonly observed themes. In particular, as set out elsewhere in this Special Report, low-carbon hydrogen projects are being developed within existing frameworks such as those for the gas, electricity and renewables sectors. This means that in many countries, low-carbon hydrogen projects are fitting into the liberalisation of electricity industries subject to independent regulators and, as such, are having to navigate a complex landscape of industry codes and legislation which was not designed with low-carbon hydrogen projects in mind.

Since low-carbon hydrogen remains a relatively new sub-sector of the industry, governments are looking to each other for guidance on how best to address the unique opportunities and challenges that the technology presents, including how to fit it into existing and evolving climate change policies. This chapter provides an overview of the key entities involved at an international and domestic level. It goes without saying that the domestic landscape is key, as beyond the policies set by the energy ministry and enforced by a regulator, investors, developers and financiers of low-carbon hydrogen projects must have a good understanding of the markets and sectors within which their specific projects will operate.

This chapter outlines the roles of the key entities but does not pretend to cover the multitude of jurisdiction-specific entities, such as regional municipalities, public–private partnership bodies, environmental agencies, planning consenting bodies, health and safety regulators, network system owners, offtakers and various stakeholders affected by low-carbon hydrogen developments.

2. Government and international cooperation

Energy has long been of both national and international interest and a key driver for economic development. For some countries it is essential to ensure they have energy independence and security, avoiding reliance upon a single producer or energy source. For others, the energy drivers may be more about minimising the impacts of climate change while maintaining economically prosperous energy markets. Some countries have fully integrated energy sectors; others prefer to maximise opportunities for competition. Either way, governments (and public institutions) play a key role in setting energy policies and strategies, be it by promoting the incumbent players (which may be partly or wholly state owned) or facilitating changes to an existing energy mix (often via subsidies and tax breaks).

Governments set and maintain their own energy vision and policy in order to make this environment sufficiently attractive for private investors. This includes developing and upholding a coherent set of laws and regulations for activities in the energy sector, and ensuring that historical provisions enable the development of new technologies such as hydrogen projects.

Another growing driver of domestic energy policy is international agreements. Ratified by 197 countries, the United Nations Framework Convention on Climate Change (the Convention or the UNFCCC),[22] which came into force in March 1994, ultimately aims to prevent 'dangerous' human interference with the climate system and is the convention under which the Paris Agreement was signed in 2015. The Paris Agreement is a legally binding international treaty on climate change whose goal is to limit global warming to well below 2°C, preferably to 1.5°C, compared to pre-industrial levels.[23]

Although coordinated, integrated policy measures are not in place at an international level, the UNFCCC framework has enabled consideration of hydrogen though measures such as nationally determined contributions (NDCs).[24] For example, in 2020 Chile

"Governments set and maintain their own energy vision and policy in order to make this environment sufficiently attractive for private investors. This includes developing and upholding a coherent set of laws and regulations for activities in the energy sector, and ensuring that historical provisions enable the development of new technologies such as hydrogen projects."

submitted an updated NDC to the Secretariat of the UNFCC that states:

> *The use of hydrogen produced from renewable energy sources for cargo transportation will play a crucial role in reaching the emissions neutrality goal, and it would be economically convenient.*[25]

Such policy signals at supra-governmental level thus serve to nudge the direction that domestic and regional policy takes.

What is clear is that governments are key to developing and promoting dedicated national policy instruments that enable the creation of low-carbon hydrogen projects, as well as ensuring that supply-and-demand-stimulating instruments are designed to allow for the rapid development of hydrogen infrastructure, along with an ecosystem of applications and end-user needs.

One example of international policy influencing individual government policy is the Clean Development Mechanism (CDM)[26] under the Kyoto Protocol.[27] This example might serve as a blueprint for government-created frameworks for low-carbon hydrogen. The CDM experience would suggest that the process can take two to three years to implement (including development, international review, revisions and approval). Using mechanisms under Articles 6.2 and 6.4 of the Paris Agreement,[28] governments can cooperate on creating emission credits, baselines and monitoring methodologies for the use of low-carbon hydrogen. These can be submitted to the supervisory body provided for in Article 6.4 to ensure a non-discriminatory, transparent and internationally verified approach. The bottom-up approach of Article 6.2 means that there is flexibility for policies to be developed to suit domestic markets and specificities while still operating within an international framework.[29]

Case study: Green Hydrogen Catapult

Launched by the United Nations in December 2020, the Green Hydrogen Catapult initiative brings together developers and partners across the globe to accelerate the scale and production of renewables-based green hydrogen.[30] This global coalition of green hydrogen project developers aims to work not only to halve the current cost of hydrogen production to below $2 per kilogram, but also to target the deployment of 25GW by 2026. In doing so, it hopes to transform the world's most carbon-intensive industries, including power generation, chemicals and steel manufacture. The aims of the initiative align with the goals of the Paris Agreement, placing hydrogen use and production on a trajectory towards net zero global emissions by 2050.

Founding partners of the project include Iberdrola, Ørsted, CWP Renewables, Envision, ACWA Power, Snam and Yara – leading private sector companies with expertise in hydrogen and confidence in its potential. The partners are working together to develop project capacity and technology, solve early market challenges and promote collaboration. The project is part of UNFCCC's Race to Zero, a global campaign to unite and mobilise actors outside of national governments, such as businesses, cities and investors, in committing to bold climate action.

International cooperation at a regional level has also been key for the setting of so-called 'hydrogen strategies', which have a central role in setting the policy in relation to low-carbon hydrogen projects – for example, the EU Hydrogen Strategy published in July 2020, which spells out why hydrogen is a key priority in achieving the European Green Deal and Europe's clean energy transition.[31] The EU Hydrogen Strategy set out a roadmap for building a low-carbon hydrogen economy in Europe in the decades up to 2050, with a phased approach for scaling up production of, and demand for, low-carbon hydrogen. This centralised, top-down approach is a shift for many investors, especially in sectors such as gas, transport and heating, where the role of government has not been as directional as in, say, the renewables sector. While this approach lends itself to creating direct policy measures such as financial support in the form of taxation (eg, of grey hydrogen) or subsidies for low-carbon hydrogen use (such as by incentivising use of hydrogen in transport), whether these additional revenues can be utilised depends on the business models that are mandated by such governmental cooperation. Furthermore, such active management of the hydrogen sector can operate to deter projects from proceeding, while investors wait and see what framework emerges.

3. Government at a national level

With low-carbon hydrogen projects trying to fit into existing regimes, whether for renewables, carbon capture and storage (CCS) or gas networks, the attitudes of governments to these sectors will drive their approach to how hydrogen is treated. A key driver is the extent to which governments have committed to addressing climate change, and particularly decarbonising. With Sweden, in 2017, becoming the first country to put into law its target to reach GHG neutrality by 2045, by December 2020 eight other countries had passed national laws committing to net zero GHG emissions over the coming decades (2050 being a popular target date).

"With low-carbon hydrogen projects trying to fit into existing regimes, whether for renewables, carbon capture and storage or gas networks, the attitudes of governments to these sectors will drive their approach to how hydrogen is treated."

The approach to promoting the development of low-carbon hydrogen is likely to follow a similar path to that adopted for renewable energy, notably wind and solar. Governments have championed the rise of renewables by creating targets, production plans and subsidy regimes; and as renewables have reached a level of maturity, policy has moved towards competitive auctions and subsidy-free regimes for new renewable projects. However, the journey for low-carbon hydrogen projects is just beginning, with government support helping low-carbon hydrogen projects compete against the much cheaper non-low-carbon hydrogen alternatives.

Case study: the Netherlands[32]

In March 2020, the Dutch government outlined its national strategy on hydrogen with a corresponding policy agenda. The strategy set out the importance of hydrogen for achieving a decarbonised energy system. The Climate Agreement, entered into between the Dutch government, industry and other stakeholders in 2019, set out ambitious targets for hydrogen, with key concepts being upscaling, cost reduction and innovation.[33] Hydrogen presents a unique opportunity for the Netherlands; it has empty North Sea gas fields that can be used for captured carbon, large offshore wind farm infrastructure with the potential to produce green hydrogen, and extensive gas infrastructure which could be used to transport hydrogen with minor adjustments. As a result, numerous hydrogen projects are in development stages, including at Groningen, recognised and receiving European subsidies as part of the first 'European Hydrogen Valley' and hosting an entire hydrogen value chain, from production and distribution to storage and local end-use.

While there is not yet any specific Dutch law in place to regulate hydrogen, the process of bringing forward legislation to enable the transportation and distribution of hydrogen by network companies has been initiated. This will be subject to a ministerial decree based on the current Gas Act. The decree is expected to clarify the role of the network companies. Legislation such as the Gas Act will also need to be updated in terms of permitting an increased percentage of hydrogen to be blended into the gas grid, particularly as the Dutch government expects hydrogen blending to reach a level of 10–20%. At present, 0.5% hydrogen is permitted in regional networks, and 0.2% in national networks, according to the Gas Quality Decree.[34]

Case study: Portugal

Portugal was the first country in the European Union to commit to carbon neutrality by 2050.[35] It is now a leader on hydrogen, with green hydrogen expected to play a significant role in helping

Portugal meet its ambitious targets, set out in the National Energy and Climate Plan 2021–2031 (NECP 2030).[36] Portugal's National Strategy for Hydrogen (EN-H2) was approved by resolution of the Council of Ministers in August 2020.[37] In addition to investment goals, it sets out further targets specific to hydrogen deployment, including, by 2030:

- a 10–15% injection of green hydrogen into natural gas networks;
- 2–2.5GW of installed electrolyser capacity; and
- 50–100 hydrogen refuelling stations.

EN-H2 will guide the gradual introduction of green hydrogen into the country's energy mix and introduce an element of incentive and stability for the future role of hydrogen. However, the regulatory framework still needs to be developed. One of the main initiatives within the strategy is to set out a regulatory framework for hydrogen. It is expected that legislation regulating the transportation, storage and reception of gas will be amended to introduce provisions relating specifically to the use and regulation of hydrogen.

The Portuguese and Dutch governments signed a memorandum of understanding to connect their 2030 hydrogen plans (especially on green hydrogen) to decarbonise the economy in line with EU climate goals.[38] The Sines project is an example of this partnership:[39] the two countries are collaborating to develop an import–export value chain for the production and transportation of green hydrogen. The Sines project will deliver green hydrogen in three ways:

- direct injection into the gas networks;
- distribution by road transport to final consumers; and
- export via the Sines port terminal.

The Sines project aims to have a capacity of 1GW by 2030 and will be powered by renewable sources, primarily solar. The Dutch connection comes from project positioning to be an exporter of green hydrogen, in particular to the Netherlands through the port of Rotterdam.

4. Sector regulator

Enshrined in international best practice and, in the case of the European Union, EU law, the role of an independent regulatory body to oversee orderly functioning of the relevant market has long been a feature of electricity and gas markets. Low-carbon hydrogen projects are likely to fit into the remit of these existing regulators, and given the increasing scale of intervention by central governments, the regulator's role is only likely to increase in the coming years.

In Great Britain,[40] the regulator is the Gas and Electricity Markets Authority (GEMA), exercising day-to-day functions through the Office of Gas and Electricity Markets (Ofgem). Currently, hydrogen (of whatever form) is defined as a gas under the Gas Act 1986 and so is subject to the same regulations that regulate the gas industry. Ofgem regulates the downstream gas industry in Great Britain, but also the electricity industry (including renewables).

Ofgem is recognised by EU directives as an independent national regulatory authority, and carries out its role in conjunction with all market stakeholders. Its powers are set out in a variety of legislation. As the national regulatory authority, Ofgem also has a role under the European Union's Regulation 1227/2011 on Wholesale Energy Market Integrity and Transparency (REMIT). In this capacity, Ofgem has a duty to uphold REMIT's framework around the prevention of market abuse in the gas and electricity markets, and is granted enforcement powers in respect of breaches by the Electricity and Gas (Market Integrity and Transparency) (Enforcement etc.) Regulations 2013. GEMA also has powers arising from relevant EU legislation that has direct effect;

"For gas and electricity markets, the traditional way that a regulator holds the operator to account is through the terms and conditions of that operator's licence."

however, these powers are under review following the United Kingdom's departure from the European Union. The regulator has the power to take enforcement action and impose significant fines for non-compliance. In the case of Ofgem, as a public body with a primary duty to protect the interests of consumers, depending on the circumstances, it can be held to account by parliamentary scrutiny, appeals to the Competition Commission, judicial review by the courts, regulatory impact assessments and consumer bodies.

For gas and electricity markets, the traditional way that a regulator holds the operator to account is through the terms and conditions of that operator's licence. Large-scale hydrogen projects will generally need a licence to produce or convey hydrogen on a network but currently are likely to benefit from an exemption when it comes to producing or supplying hydrogen to customers. Whether low-carbon hydrogen projects require a licence from the regulator depends on the application and activity.

Unless an exemption applies, the regulator's role will be to assess the licence application, and to issue, modify, enforce and revoke relevant licences. Although most licences are standard and in practice are largely concerned with the ability of the hydrogen operator to comply with existing industry codes, the challenge for low-carbon hydrogen projects is that the industry codes do not specifically regulate low-carbon hydrogen and often the standards and guidance on which the industry relies have methane gas rather than hydrogen in mind. Working out what applies and how to ensure that a low-carbon hydrogen project is not in breach of its regulatory obligations and/or is not held back by the existing regime, is a role the regulator will need to embrace.

For example, transportation of hydrogen (of whatever type) generally requires a gas shipper licence under section 7(3A) of the Gas Act 1986 to convey gas through pipes. However, due to competition rules deriving from EU 'unbundling' regulations, a company that holds a gas shipper licence must not also hold a gas producer licence (unless an exception applies). This creates a barrier to the same company producing and conveying low-carbon hydrogen, which the regulator alone cannot resolve.

Transport does not fit within Ofgem's remit, which creates a level of uncertainty for low-carbon hydrogen projects operating in the transport sector. To the frustration of some developers, while Ofgem has extensive experience and expertise in coordinating and monitoring various regulatory regimes which encourage new technologies, such as use of renewable heat, solar and wind subsidies (though most of these have now ended), the regulation of the transport sector lags behind and does not coordinate well with the gas and electricity sectors. This situation is unfortunately common to a number of jurisdictions where the historical development and separation of government ministries has led to divergence in approaches, and it is already having an adverse impact on low-carbon hydrogen transport project development (see chapter IV).

5. Producers

As set out in chapter III, there are multiple ways in which low-carbon hydrogen may be produced at present. When it comes to hydrogen, the two dominant production methods are:

- electrolysis, using renewable electricity to make green hydrogen; and
- steam methane reformation, using natural gas (methane) with CCS to make blue hydrogen.

Although they are both low carbon, green and blue hydrogen attract companies with differing interests.

5.1 Green hydrogen

For green hydrogen, the producers have typically had some exposure to the renewables sector and possibly other storage technologies such as hydro plant and/or lithium-ion batteries. This is because green hydrogen production requires an understanding of how the electricity market, and in particular the renewables sector, in the given jurisdiction works, in order to operate an electrolyser project within it. For example, the rights required to obtain relevant consent; land rights; and the dependability of supply of electricity inputs – all would be key for such a project in order to structure the necessary offtake arrangements. As further described in chapter VI, the contractual structures will take into account grant and concession arrangements with any state or quasi-state entity, reliability of electricity input arrangements, and requirements for a stable quantity of hydrogen offtake.

Where the green hydrogen requires transportation and storage, for example to be used in transport fuelling at a different location, the developer would also need to comply with the health and safety requirements of the relevant jurisdiction. For example, in the United Kingdom the Health and Safety at Work Act 1974 applies to work activities, including hydrogen projects. The act requires employers to ensure that the health, safety, and welfare of their employees is not put at risk by the conduct of their business. Because hydrogen is classified as a dangerous gas, additional health and safety measures must be observed. In short, the health and safety legislation aims to ensure that organisations carry out their commercial activities in such a way as not – insofar as is reasonably practicable – to expose other people to risks to their health and safety. This means that the entity responsible for enforcing health and safety law (in the United Kingdom the Health and Safety Executive (HSE)), plays a key role in ensuring that health and safety matters are carefully managed, monitored and complied with.

5.2 Blue hydrogen

For blue hydrogen, producers have been companies with an interest in developing CCS technologies, and so comprise a more diverse group. Because CO_2 captured through CCS is also a gas, like hydrogen, for such producers the structure of the gas market and management of the gas network are of primary interest.

As discussed further in chapter V, the availability and regulation of network infrastructure to inject and offtake hydrogen on the gas network, including the relevant charges and obligations, will be the key aspects for blue hydrogen developers. For such producers, the questions of network capacity, and flexibility with regards to gas quality, are the areas to consider because many countries do not at present permit the injection of hydrogen onto the networks. Further, the injection and offtake of hydrogen at different points on the network will require a change to the historic use of the gas system, so producers of blue hydrogen in particular will need to work closely with the gas network operators to actively manage both the network and the quality of gas.

6. Consumers

Consumers of low-carbon hydrogen are what make up the so-called 'demand' which producers are keen to marry up with their 'supply'. Currently the challenge for many producers is the ability to find and then satisfy the demand.

Consumers can be broadly divided into two categories:

- large business and industrial; and
- end consumer.

At the industrial end, the value of low-carbon hydrogen would lie in the ability to decarbonise sectors such as manufacturing (cement, glass, steel etc); whereas at the consumer end it's mainly about decarbonising transport (fleet vehicles, public transport etc) and heating and cooling in residential and commercial buildings (via hydrogen boilers and hydrogen fuel cells). In any case, as set out in Figure 3, the annual hydrogen consumption (in this example, in the United Kingdom) is due to rise even with modest efforts to decarbonise some fundamental areas such as the service, industry, transport and residential sectors.

Figure 3. Annual UK hydrogen consumption by sector in 2050, for three scenarios with net zero GHG emission targets

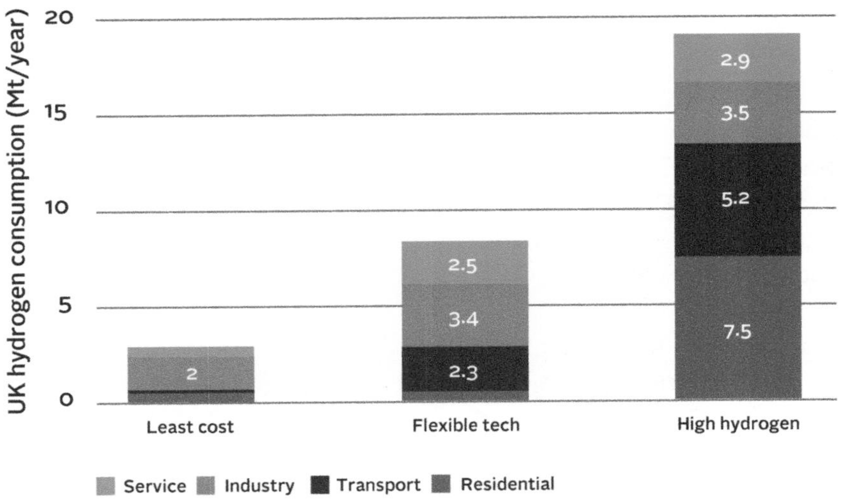

Source: Dodds et al, "Opportunities for hydrogen".[41]

In each case the main opportunity for developing a hydrogen economy is ensuring that such consumers choose low-carbon hydrogen over any other option that may be available to them. For this, the need for reliable production at competitive cost has been the challenge to date. Although IEA analysis finds that "the cost of producing hydrogen from renewable electricity could fall 30% by 2030 as a result of declining costs of renewables and the scaling up of hydrogen production",[42] according to a 2020 report by Capgemini Invent and Breakthrough Energy, "at around €6 per kg, green hydrogen is not competitive today with fossil energies (parity at €1/kg equivalent to €25/MWh)".[43]

As noted above, the role that governments play in encouraging low-carbon hydrogen consumption will impact consumer behaviour. For example, at the EU and UK levels, governments are considering modifying obligations on fuel suppliers to increase the percentage of renewable fuels (of which green hydrogen is one) in transport. Failure to do so would result in financial penalties for the fuel supplier, the regime being similar to that in place for renewables in the United Kingdom under the Renewables Obligation (RO) model.[44] The additional revenue raised would be passed back to producers, thus creating a revenue stream for low-carbon hydrogen.

In the industrial sector, where consumers are more susceptible to global competition and prices, this may be more challenging. Consumers may need further encouragement through grants, tax breaks and other subsidies, as well as public pressure from downstream consumers for low-carbon hydrogen-fuelled products. For example, one way low-carbon hydrogen could have an impact is in attracting high rents for buildings with lower environmental impact. Surveys have consistently indicated that office tenants are willing to pay more for green features – a survey of 708 respondents in 17 major US markets from 2016 revealed that office tenants would be willing to pay a 9.3% premium.[45] Such statistics have only accelerated with the rise of the environmental, social and governance (ESG) agenda and concerns around air quality. In a 2021 survey of 1,500 occupiers by international law firm CMS, 65% of respondents said they would happily take a pay cut to work in a sustainable building.[46] Such purchasing power and consumer pressure is critical for the development of the hydrogen sector and market.

7. Finance and insurance

To date, several governments have supported research and development of low-carbon hydrogen projects. For example, in Australia, the government-backed Clean Energy Finance Corporation created a US$300 million Advancing Hydrogen Fund[47] and the

"The role that governments play in encouraging low-carbon hydrogen consumption will impact consumer behaviour. For example, at the EU and UK levels, governments are considering modifying obligations on fuel suppliers to increase the percentage of renewable fuels (of which green hydrogen is one) in transport."

Australia Renewable Energy Agency (ARENA) held a US $70 million Renewable Hydrogen Deployment Funding Round at the federal level.[48]

However, while public sector grants and subsidies will have a role to play, the contribution of the private sector cannot be underestimated. Longer-term, large-scale hydrogen production and the need for significant infrastructure for blue hydrogen projects, given the cost disparity of all low-carbon hydrogen compared to current, predominantly grey, hydrogen production methods, mean that the role investors and financiers play in the nascent sector will be critical to ensuring the transition from pilot projects and grant-based initiatives to large-scale commercialisation. For example, according to Hydrogen Europe's Green Hydrogen Investment and Support Report, to achieve the hydrogen ambitions set out by the European Union would require an estimated €430 billion of investment, of which €120 billion would be for hydrogen infrastructure and storage.[49]

As discussed in chapter VI, the willingness of the private sector to invest funds in low-carbon hydrogen projects will depend on the legal and policy landscape of the jurisdiction in which such projects would operate.

Supra-national influences

Governments can influence the development of hydrogen projects at a supra-national level in a number of ways:

- **Encouraging international cooperation** through political initiatives, coordinating research and development and removing barriers to development of projects. This could be through cooperation councils and/or dedicated institutions responsible for policy coordination, echoing the role played by IRENA for renewable energy.
- **Supporting development** of the long-term low-carbon hydrogen infrastructure needed to generate and transport hydrogen across international borders; or from areas that are seen as likely suppliers of low-carbon hydrogen, such as Australia, Canada and the USA, to countries that are likely to be key consumers, such as Japan and the European Union.
- **Creating international standards** across the value chain for standards monitoring, reporting and verification (MRV) requirements, as well as safety standards for production, transportation and storage of hydrogen.
- **Creating incentives** for both producers and end consumers to switch to low-carbon hydrogen. This is likely to include:
  - developing and testing financial incentives for generating emission credits under international market mechanisms;

- creating low-carbon hydrogen 'feed-in' tariffs (ie, price premiums similar to renewable feed-in tariffs);
- tax breaks for low-carbon hydrogen projects that stay below certain emission limits; and
- other subsidies that encourage uptake by covering the price disparity between grey and low-carbon hydrogen.

- **Developing long-term economic partnerships** including through the use of bilateral initiatives under the cooperative approaches of Article 6.2 of the Paris Agreement and building on previous initiatives such as those undertaken in the context of the Joint Crediting Mechanism (JCM) for bilateral collaboration on GHG mitigation.[50] Several EU member states have expressed their interest in developing bilateral agreements under the UNFCCC, and others such as Saudi Arabia, Japan and China would be well suited to similar regional developments.

"Green hydrogen is projected to grow rapidly in the coming years and is hydrogen's answer to decarbonisation. While technically viable today the key is for it to achieve economic competitiveness, and it is heading in that direction."

III. Production

1. Types of hydrogen

Hydrogen on its own is the most abundant element on earth, estimated to contribute 75% of the mass of the universe. It rarely exists in its elemental form, but in compounds contained in most molecules of living things. On its own, this gas would be odourless and colourless. However, in the context of its use as an energy carrier, a 'rainbow' of hydrogen colours has developed. These colour references are to the energy source and/or process used to produce that type, or 'colour', of hydrogen when extracting it from its compound state. A short glossary is set out in Table 1, below. That said, the definitions of the different types of hydrogen are evolving and, once these definitions become settled (ie, through market practice or in law), will have legal implications for stakeholders. For example, with the EU taxonomy published in March 2020,[51] 'low carbon' or 'renewable' hydrogen is limited to blue and green colours by virtue of requirements that:

- direct CO_2 emissions from manufacture of hydrogen are limited to 5.8 tCO_2e/t;
- electricity use for hydrogen produced by electrolysis is at or lower than 58 MWh/t; and
- average carbon intensity of the electricity produced that is used for hydrogen manufacturing is at or below 100 gCO_2e/kWh.

Broadly, in this Special Report we will talk about the three main types of hydrogen:

- **Green:** Green hydrogen is produced by electrolysis, the process of splitting up water (H_2O) into hydrogen and oxygen. So long as the electricity used for the process of electrolysis is renewable (ie, solar, wind etc) it is 'green'. For sectors that are difficult to decarbonise, such as industrial processes, heavy transport and aviation, green hydrogen is a good alternative where solar and wind are not feasible. Green hydrogen is projected to grow rapidly in the coming years and is hydrogen's answer to decarbonisation. While technically viable today the key is for it to achieve economic competitiveness, and it is heading in that direction.
- **Blue:** Blue hydrogen is seen as carbon neutral but not renewable, as it involves the use of gas combined with carbon capture technology. Blue hydrogen is seen as an intermediate step, as up to 90% of the CO_2 emissions released are captured during production and stored.

For purposes of this Special Report, the term 'low-carbon hydrogen' is used to refer to both green and blue hydrogen.

"For sectors that are difficult to decarbonise, such as industrial processes, heavy transport and aviation, green hydrogen is a good alternative where solar and wind are not feasible."

- **Grey:** Grey hydrogen involves extracting hydrogen out of natural gas or coal. Currently 90–95% of the hydrogen produced in the world is grey. At present, grey hydrogen is mainly produced by reforming natural gas for industrial applications. This process releases a lot of CO_2. The main applications are in the chemical industry for making ammonia and fertiliser, and for oil refining.

Case study: offshore green hydrogen production – Oyster project[52]

At the beginning of 2021, a public–private partnership of the European Commission, the Fuel Cells and Hydrogen Joint Undertaking,[53] awarded €5 million to a consortium of companies investigating the feasibility of producing green hydrogen at offshore wind farms in Europe. By combining offshore wind turbines directly with an electrolyser, the project hopes to demonstrate an ability to transport green hydrogen to shore.

The project, named 'Oyster', is being developed by ITM Power and Ørsted, along with Siemens Gamesa Renewable Energy and Element Energy, and is anticipated to run to the end of 2024. It will develop and test a megawatt-scale fully 'marinised' electrolyser in a shoreside pilot trial. The electrolyser system will be compactly integrated within an offshore wind turbine and use seawater as a feedstock for electrolysis by means of an integrated desalination and water treatment process. The project aims to harness the potential of offshore wind and develop an electrolysis system that can withstand harsh offshore environments with minimal maintenance, while meeting cost and performance targets. Combined with a sufficiently high carbon tax, the consortium believes hydrogen produced in this way can be cost-competitive with natural gas, and therefore presents significant potential to facilitate the transition to fully renewable energy systems in Europe.

As the research and development of hydrogen continues, more 'colours' of hydrogen are discovered and developed, so watch this space.

Table 1. Glossary of hydrogen 'colours'[54]

Colour	Explanation
Grey	Hydrogen generated using methane gas from steam reforming process. This is by far the most common way of producing hydrogen, accounting for about 94% of all hydrogen production and releasing about 9.3kg of CO_2 per kg of hydrogen produced. The hydrogen produced costs between $1 and $3 per kg.
Brown/black	Hydrogen generated from gasification of coal to gas. This is the oldest and most polluting way of producing hydrogen (it releases CO_2 and carbon monoxide to the atmosphere). The colour indicates what type of coal is used: brown (lignite) or black (bituminous) coal. The hydrogen produced costs between $1 and $2 per kg.
Green	Hydrogen generated through electrolysis process using electricity generated from renewable sources. This is the ultimate aim of many low-carbon hydrogen policy makers, but is behind blue hydrogen in terms of being a low-carbon hydrogen-making process achievable at scale. At present, it accounts for about 1% of overall hydrogen production. The hydrogen produced costs between $3 and $7.50 per kg.
Blue	Hydrogen generated by methane gas from steam reforming process where the associated CO_2 is captured and stored underground via CCUS technologies. This is the low-carbon hydrogen expected to enable the scaling-up and commercial roll-out of hydrogen at scale. The hydrogen produced costs between $1.50 and $3 per kg.
Pink	Hydrogen generated by the electrolysis process using electricity generated from only nuclear generation. The cost of hydrogen produced using this method varies a lot depending on the characteristics of the nuclear generation involved.
Turquoise	Hydrogen generated using methane gas by using the molten metal pyrolysis process. This method is currently in experimentation stages so is not commonly referred to.
Yellow	Hydrogen generated through electrolysis using electricity generated from only solar generation. This colour is not commonly referred to.

2. Regulation

While at present most countries do not explicitly regulate how hydrogen is produced, and thus do not incentivise the choice of hydrogen production method, 2020 has seen a number of policy roadmaps which indicate that green and blue hydrogen are the favoured colours. However, the big challenge for any producer or user of hydrogen is the cost differential and the lack of necessary infrastructure needed to make the jump from grey to low-carbon hydrogen alternatives.

Cost is the primary barrier for making the switch between grey and low-carbon hydrogen. According to analysis carried out by the economic consultancy firm Pöyry, by 2050 production costs of blue and green hydrogen will still be above those of grey (see Table 2). And according to the Oxford Institute for Energy Studies, even if renewable electricity for use with electrolysis falls to EUR30/MWh, the green hydrogen cost would be in the range €50–60/MWh.[55] Blue hydrogen costs are likely to be lower (and at similar efficiency levels as green hydrogen) and, at least initially, likely to allow the production of hydrogen on a greater scale than through electrolysis.

Table 2. Projected comparison of low-carbon production cost and efficiency for 2050[56]

	Capex (€/kW H2)	Opex (€/kW H2)	Efficiency %	Levelised cost (€/MWh)
Blue hydrogen	934	37	78	47
Green hydrogen	544	31	80	66

On that basis, production of hydrogen may require additional incentives to overcome the cost barrier. This may be by way of carbon taxes, tax breaks, or other subsidies, while additional incentives are needed to promote the construction of the additional infrastructure needed to deploy low-carbon hydrogen (such as the CCUS infrastructure needed for blue hydrogen).

However, hydrogen has centre stage from a policy perspective in a number of jurisdictions. For example, the European Commission has set a clear direction for the type of hydrogen to be produced in the EU.[57] While insufficient in isolation, such policy directions, especially when made in collaboration with the industry and other stakeholders, have the potential to create the necessary regulatory and legal frameworks that give effect to the policy aims of the relevant jurisdictions.

Some of this is already emerging, for example in some EU member states. However, the policy is piecemeal and the lack of a joined-up legislative framework is an evident barrier. While on the one hand the European Union's hydrogen strategy encourages the uptake of green hydrogen production, the legal framework for the transport and storage of hydrogen has not caught up with this aim. The existing Gas Directive[58] focuses on competition between the entities performing different roles within the gas network, but makes little provision for an increase in hydrogen on the networks and is silent on the crossovers between the electricity and gas networks which an energy carrier like hydrogen would introduce.

Furthermore, hydrogen production does not fit neatly into current schemes which are available to other methods of low-carbon energy use or production. For example, many countries have a system of guarantees of origin which track whether the electricity is renewable. Emanating from the European Union's Renewables Directive,[59] this system includes within it a mechanism such that guarantees of origin may be issued for renewable gases (including green hydrogen). This not only allows final consumers to know that the energy consumed is from renewable sources, but, as suppliers place a premium on renewable energy, provides an increased revenue stream for the producers. However, given that the system was designed for wind and solar PV electricity generation, a producer of green (or blue) hydrogen has to navigate a regime that doesn't completely fit its characteristics. As discussed further in chapter V, the traceability of hydrogen and lack of an established market create barriers not faced by renewable technologies or applicable to grey hydrogen.

IV. Transportation and storage

1. Introduction

In most jurisdictions, legislation specific to the transportation and storage of hydrogen or the operation of hydrogen networks has not been developed. Instead, the relevant legislation relating to the regulation of gas is expected to govern the transmission, distribution and storage of hydrogen.

While that may not always be the case (see the case study below in relation to proposals for gas and hydrogen regulation in Germany, for example), current hydrogen developments are operating within the context of existing gas regulations. In the United Kingdom, for example, the Gas Act 1986 gives Ofgem a duty to regulate and protect the interests of consumers with regard to persons or entities engaged in, or connected with, "the shipping, transportation or supply of gas conveyed through pipes".[60] Any entity engaging in gas supply, gas shipping or gas transportation, or which participates in the operation of gas interconnectors or provides smart metering in respect of gas, must have obtained a licence to do so under the Gas Act. The licences include measures relating to the safe operation of the gas network and to price controls, and licensed entities must demonstrate a credible plan to commence licensed activities and permit a risk assessment to be carried out by the regulator.

Additionally, across most EU states, the International Carriage of Dangerous Goods by Road (ADR) will have been implemented which will regulate how hydrogen, and other dangerous gases, are to be transported between countries. Under Annex 5 of the ADR, hydrogen is classified as a dangerous good, and any vehicles carrying hydrogen must be of a certain specification and drivers appropriately trained.

Case study: should methane and hydrogen be regulated separately?

In February 2021, the German government passed a draft amendment to the Energy Act with the aim of gradually building up a 'pure' hydrogen infrastructure in Germany (ie, pipelines that carry only hydrogen and not other gases), but regulating it separately from the existing gas (methane) pipeline network.[61] This would mean that existing planning rules for the gas grid and how it is financed by way of grid fees would not automatically apply to pipelines carrying hydrogen. Currently, networks transporting only hydrogen are not regulated under German law as they do not fall under the heading of 'energy' in the Energy Industry Act. The amendment would widen the definition of 'energy' (at s 3 no 14 of the Act) to include hydrogen, so that it would become a third, independent, designated source of energy, after electricity and gas. However, this would not apply to the blending of hydrogen into the methane gas network, so the existing legal framework would continue to apply.

The bill will allow operators of hydrogen networks to choose whether they want to be subject to the new regulation, and any benefits and restrictions that come with it. This way, the legislation will aim to facilitate the development of hydrogen networks without hindering those networks carrying hydrogen and natural gas blends. However, because network operators are required to grant access and connection to their hydrogen networks based on the principle of negotiated network access, if hydrogen networks are to be regulated separately the industry standard contracts for regulated network access to the natural gas network, which have been in continuous development since 2006, would be up for review. Similarly, the conditions of network access, and the tariffs that would apply, are not adequately addressed by the proposed amendments. It is not clear, for example, to what extent the existing Ordinance on the Incentive Regulation of Energy Supply Networks, which governs the conditions and tariffs for network access (noting that these must be reasonable, non-discriminatory and transparent), could be reopened.

Thus far, the proposed separation in regulation of the methane and hydrogen networks has met with strong opposition from the industry and market participants. Amongst the reasons cited are that developing and testing new regulation piecemeal will slow down rather than accelerate the development of hydrogen networks and will prevent the coordinated development of gas and hydrogen infrastructures as more hydrogen is added to the gas networks.

2. Health and safety issues

As it is primarily regulated as an explosive gas, hydrogen, like other gases, will likely also be heavily regulated from a health and safety perspective. Often, the regulator will not only restrict the storage of large quantities of gases such as hydrogen but will also, as part of gas quality standards, mandate parameters as to the amount of hydrogen that can be included on the methane gas networks, the pressure at which any gas is transported and further restrictions to ensure that the gas is handled in a safe manner. Such restrictions may be driven by health and safety requirements but may also be a result of market practice that reflects national and regional characteristics, regularised by way of industry codes that apply to the gas market in the particular jurisdiction. The industry codes work as multilateral agreements between stakeholders that define the technical and commercial terms within which they must operate. These are then given legal footing by way of the licence conditions with which a gas licensee would be required to comply.

These are some examples of the regulations with which the United Kingdom's HSE requires compliance:

- **Gas Safety (Management) Regulations 1996:** These concern the flow of gas through the network. All gas transporters must prepare and submit a safety case to HSE. This identifies the hazards and risks, explains how they are controlled and describes the system in place to ensure that controls are applied. The gas transporter will be audited to ensure compliance with their safety case.
- **Pipeline Safety Regulations (1996):** These are about pipeline integrity – requirements in respect of pipeline design, construction, installation, operation, maintenance and decommissioning. For example, pipelines should be equipped with emergency shut-down valves and their design should take account of the need for maintenance access.
- **Planning (Hazardous Substances) Regulations 2015 and/or Control of Major Accident Hazards Regulations 2015 (COMAH):** These regulate storage of hydrogen, depending on the quantities involved. COMAH sets a high bar, requiring operators to take all measures necessary to prevent a major accident and limit consequences for human health and the environment. The

"One lesson from the 'new' technologies developed over the last few decades, including renewables, has been that familiarity with the authorities and stakeholders responsible for granting property and consenting rights is a key consideration."

operator must have in place various strategies, including safety and emergency plans and a major accident prevention policy.

- **Hazardous Substances Regulations:** Consent is required for storage of two or more, and further consent for five or more, tonnes of hydrogen.
- **Dangerous Substances and Explosive Atmosphere Regulations 2002:** These set out requirements for the use of equipment and protective systems in potentially hazardous environments, including those where hydrogen is produced or stored.
- **Pressure Equipment (Safety) Regulations:** These regulations apply to the design and manufacture of tanks that will be used for the transportation of hydrogen.

In addition to legislation, there are a number of industry codes that contain requirements for gas licensees:

- **The Uniform Network Code:** This sets out common rules governing the gas transportation arrangements between licensed gas transporters and shippers, as required under their licence. Every licensed gas transporter must have its own network code, incorporating the Uniform Network Code and governing the terms on which it will transport gas. It includes a transportation principal document (TPD), which sets out the gas transportation arrangements between gas shippers and transporters, and an offtake arrangements document which sets out arrangements between different transporters.

- **Independent Gas Transporter Uniform Network Code:** This sets out the common rules applying to independent gas transporters, who operate extensions to the gas network such as those serving new housing developments.
- **Supply Point Administration Agreement:** This is a multi-party agreement with which all gas transporters and suppliers are required to comply. It facilitates supply point – change of gas supplier – administration.
- **Retail Energy Code:** This enables end consumers to switch energy suppliers.

3. Consents and land rights

Chief among legal considerations for hydrogen transport and storage is the overall framework, and the ability to secure the necessary authorisations and land rights. Given the treatment of hydrogen within the gas network framework, permissions and land-related rights are often interrelated with the network requirements but still tend to be rooted in the requirements and history of a particular location. In jurisdictions with a well-established consenting regime that lends itself to the development of hydrogen projects, the process is likely to be much easier than in jurisdictions where the process may be more opaque and piecemeal. In particular, for major projects the relevant jurisdiction may have a specialised department dedicated to making such decisions, or these may be made at ministerial level, such as by the Ministry of Energy, to ensure that infrastructure, land and other related decisions are made in a coordinated and comprehensive manner.

Major hydrogen projects will almost certainly require the construction of new, or the development of existing, national infrastructure. The consenting regimes of each country are localised to the particular needs of that jurisdiction. Further, projects will often need to ensure that the application for any infrastructure development takes into account the views of other stakeholders. The level of scrutiny imposed by the consenting regime may mean that securing land rights and relevant consents is time consuming, contributing to the lengthy timescales of hydrogen project developments.

One lesson from the 'new' technologies developed over the last few decades, including renewables, has been that familiarity with the authorities and stakeholders responsible for granting property and consenting rights is a key consideration. At least initially, the lack of clear guidance and precedents for these technologies introduced additional delays and uncertainties into the process, as local jurisdictions developed their practices for dealing with what were then projects without precedent. Some of the lessons learned can be applied to hydrogen projects.

In addition to ensuring the relevant consents are obtained, as with other infrastructure types, land rights need to be secured. Access rights would be needed for production and storage facilities to ensure they meet the requirements of large-scale industrial transportation, either through private contracts or under compulsory acquisition powers. In the case of repurposing existing infrastructure, variations to existing rights are likely to be needed to reflect the necessary technological upgrades and/or regulatory issues.

In a number of jurisdictions, an environmental impact assessment will be required if hydrogen is to be stored on site or if there are pipelines carrying hydrogen. How stringent these requirements are will depend on the relevant local laws.

4. International transportation

Gas pipelines have long run across national boundaries and several regions have developed a complex web of gas networks and pipelines which aim to address concerns around price to consumers and energy security. As governments and gas network operators consider the long-term impacts of increasing amounts of hydrogen in the gas networks, the ability to transport hydrogen internationally and to store hydrogen on the network are being considered.

One such example of regional cooperation is the European Hydrogen Backbone.[62] This is an initiative of 23 Europe-based gas infrastructure companies and transmission system operators (TSOs) (including Enagas, Gasunie and Snam), which seeks to connect up existing gas pipeline infrastructure to create a pan-European network. The network – or 'backbone' – will link hydrogen supply with demand, predominantly feeding European cities with hydrogen produced from offshore wind or solar PV. To begin with, the hydrogen backbone will connect 21 European countries via 11,600 km of pipeline by 2030, stretching to 39,700 km by 2040. It is expected that 69% of the pipelines used by 2040 will be repurposed existing pipeline and, on this basis, total investment could be up to €81 billion. Figure 4 shows the proposed pipeline routes across western and central Europe.

Figure 4. Proposed European Hydrogen Backbone[63]

Source: Reproduced with permission from Guidehouse (see note 63).

The proposal also outlines changes that will need to be made for the existing pipeline infrastructure to be capable of transmitting varying blends of hydrogen. However, the necessary modifications are not as significant as first imagined. Following research and pilot projects undertaken by European gas TSOs, existing natural gas pipeline materials were found to be "generally fit for hydrogen transport".[64] Largely, the proposed amendments consist of reconstructing damaged or cracked pipelines and replacing valves close to the end of their life expectancy, to ensure minimal hydrogen leakage.

Case study: Asia-Pacific hydrogen supply chain

In October 2020, the Japanese Industry Minister announced Japan's intention to establish a commercial hydrogen fuel supply chain by 2030.

The project, named HySTRA, is the product of a consortium of seven companies (J Power, Shell Japan, Iwatani Corporation, Kawasaki Heavy Industries, Marubeni Corporation, ENEOS Corporation and K Line) and aims to establish a hydrogen energy supply chain pilot project between Australia and Japan:

The association [is] working towards creating a CO2 free hydrogen energy supply chain comprised of hydrogen production effectively utilizing brown coal, transportation, storage and utilisation of hydrogen, and establishing and demonstrating the technologies to commercialise the supply chain around 2030.[65]

The pilot will produce hydrogen from brown coal at a gasification plant in the Latrobe Valley, before transporting it 150 kilometres across country, where the hydrogen will be refined, liquefied and stored on site at a port in Hastings, Melbourne. The hydrogen will then be shipped approximately 9,000 kilometres to Kobe Airport Island, Japan, for its end use. The pilot began in early 2021 with different companies from the consortium leading on different aspects of the supply chain. For example, J Power are overseeing the production of hydrogen at the gasification factory, whereas Shell Japan, Kawasaki and Iwatani will lead on the liquefied hydrogen carrier and storage and unloading facilities.

5. Issues and opportunities

Inevitably, modifications to existing codes and legislation specific to the transportation and storage of hydrogen – or the introduction of new, hydrogen-specific codes – will be frequent and essential. This is particularly true where primary legislation and industry codes only account for hydrogen in the context of its use for 'traditional' purposes, for example, grey hydrogen used as industrial feedstock or in oil refineries, rather than as an energy source. This is currently the case in China, France and Romania; and in Spain, electrolysis is classified as an energy use rather than an energy conversion process, subjecting green hydrogen to additional charges and environmental constraints.

Furthermore, where the calorific content of hydrogen is altered through the blending of hydrogen and natural gas (methane), new industry codes may need to be introduced to govern the varying content of hydrogen within the gas network. In Great Britain, for example, amendments could be made to the existing Uniform Network Code's TPD to allow for hydrogen entering the national network. Key amendments may include updating system entry and exit requirements to differentiate between hydrogen connected to the network and hydrogen conveyed between particular entry and exit points in a non-linear fashion. Further, a dedicated hydrogen network entry agreement and applicable offtake arrangements may need to be introduced to ensure hydrogen producers and offtakers are also bound to the operational framework of the TDP. And a hydrogen-specific pipeline system may need to be created alongside the national transmission system (NTS) and local distribution zones (LDZs), depending on where hydrogen will be distributed. Finally, the classification of system entry points, system exit points, storage connection points and supply points may also need to be altered to include hydrogen-specific points across the network.

Another concern arising from the way the European legal landscape has developed is around the concept of 'unbundling'. This is the principle that network industries, such as gas, separate the activities in the value chain that are potentially subject to competition (such as production and supply of gas) from those where competition is not possible or allowed (such as transmission and distribution of gas, which operate as regulated monopolies). Determining the adequate level of unbundling required in a network has been the subject of extensive discussion and development across the European Union and follows a set of permitted regulatory models. These rules would apply to hydrogen transmission and distribution networks on the same basis that they apply to existing methane gas networks. Indeed, proposed amendments to German energy law expressly legislate for unbundling: section 28m of the proposed German Energy Act would stipulate that hydrogen network operators may not construct, operate or own facilities for the production, storage or distribution of hydrogen. While the principle is a cornerstone of European competition regulation, its application to a nascent network like that for transmission and distribution of hydrogen is not without its challenges.

V. Applications

1. Industrial processes

1.1 Introduction

There is already a significant demand for hydrogen in industrial processes, and here we consider some of the key legal and regulatory issues for low-carbon hydrogen applications in the industrial sector.

The IEA estimates that total global demand for hydrogen will be around 40 million tonnes per year over the coming decade.[66] This includes using hydrogen in various industrial processes, for example oil refining (33% of all hydrogen produced currently is used in oil refining), ammonia production (27%), and steel production (3%).[67]

Nearly all industries will consume natural gas in some form, whether through feedstock or in heating processes. The hydrogen used in these industries is generally grey, and as such there is a large opportunity to decarbonise the sector by transitioning to blue or green hydrogen. This is particularly the case with blue hydrogen as carbon capture technology continues to be developed.

As a result, many major projects are being undertaken to decarbonise industrial processes. Swedish state-owned energy company Vattenfall, for example, is involved in a number of hydrogen projects that aim to decarbonise the sector. In the Netherlands, the company is collaborating with Equinor and Gasunie in an ongoing project that aims to convert a gas-fired power station to hydrogen.[68] In Sweden, Vattenfall is collaborating with steel producer SSAB and mining company LKAB on its HYBRIT project, working on emission-free steel production by replacing coking coal with hydrogen. If successful, and implemented at full scale, this project alone could reduce Sweden's CO_2 emissions by 10% and Finland's by 7%.[69]

1.2 Development of hydrogen industrial clusters

Historically, industrial processes have developed in fairly localised industrial groupings or geographical clusters. In view of this, and the infrastructure needed to service industrial processes using hydrogen, a number of countries are planning to develop new hydrogen projects focused on the needs of such industrial clusters. One example is the HyNet net zero carbon project in the United Kingdom, which proposes a hydrogen cluster in which blue hydrogen is produced to supply a range of industrial sites in northwest England, with the carbon emissions of these sites directly captured.[70] Another is Net Zero Teesside, a full-chain CCUS project, which aims to decarbonise a cluster of carbon-intensive businesses by as early as 2030.[71] The location on the River Tees was partly chosen because industries in the Teesside region account for approximately 5.6% of the United Kingdom's industrial emissions and the location offers access to storage sites in the southern North Sea, with the potential to reach over a gigatonne of CO_2 storage capacity.

Similar hydrogen industrial clusters are being considered also in the Netherlands and Belgium. Smart Delta Resources (SDR) is a cross-industry consortium of "energy and resource-intensive companies" throughout the Netherlands and Flanders, which aims to operate as a taskforce driving a CO_2-neutral industry throughout the region. The Schelde-Delta region, which is hub for large chemical, energy, steel and food corporations, is a heavily carbon-emitting area (currently responsible for emitting 22 Mt of CO_2 a year). It also already accounts for the largest regional consumption of hydrogen in Benelux, at around 580 kt of hydrogen a year. SDR has created the Hydrogen Delta Program with the aim of phasing out grey hydrogen in the region, through large-scale blue and green hydrogen production, as well as yellow (solar) and pink (nuclear) hydrogen.[72]

A key advantage of developing hydrogen clusters is the immediate link between production and demand without the need for extensive new infrastructure. If clusters can then be interconnected, as is being proposed in the Netherlands, this could create efficiencies and networks of shared infrastructure – particularly useful where there are industrial activities at a shared location, such as at ports. For example,

"A key advantage of developing hydrogen clusters is the immediate link between production and demand without the need for extensive new infrastructure."

at Porthos in the port of Rotterdam, the intention is to capture CO_2 from the port, and the port of Amsterdam, and allow for further sharing of infrastructure for the benefit of all users.[73]

In this context, the key legal and regulatory issues associated with the use of low-carbon hydrogen in industrial processes can be grouped into three broad areas:

- health and safety considerations;
- land rights and consenting regimes; and
- opportunities for revenue maximisation.

Given the early stage of hydrogen developments, this Special Report does not address issues relating to decommissioning of hydrogen infrastructure and/or the wider environmental impacts that may arise in the context of wider on-site chemical usage.

1.3 Health and safety (and technical) issues

As discussed elsewhere in this Special Report, research is being undertaken as to the safety of hydrogen blending within existing gas networks. An additional challenge in servicing carbon-intensive industrial processes with hydrogen through existing networks would be the need to produce and deliver the hydrogen at scale. Economic feasibility studies are underway in a number of countries, led chiefly by the gas transmission operators. In the United Kingdom, for example, National Grid Gas – the GB gas transmission owner and system operator – is leading research into the ability to blend hydrogen with methane on the existing gas network. Now that the technologies have been selected, the focus has shifted to network safety trials, with the aim of understanding the possible safety implications of transporting hydrogen in high quantities to industrial offtakers.[74] Once such trials have been completed, hydrogen blending can be tested against use in current industrial processes as part of future considerations. Clearly the success of blending more hydrogen onto the gas networks with the aim of offtaking for use in industrial processes will also depend on the commercial factors involved. At present, the switch from grey hydrogen to any low-carbon hydrogen option is economically challenging.

1.4 Consent issues

As discussed in chapter IV, hydrogen transportation is currently largely regulated under the existing gas legislation. As such, its use within industrial processes is often authorised on the basis that the hydrogen is likely to be produced on site and used as part of established industrial processes. Changing this set-up to include the ability to

"There are currently few public financing mechanisms for the hydrogen industrial market. Instead, many of the available financing options are a mix of public–private funding."

pipe in low-carbon hydrogen and/or include the ability to produce hydrogen closer to the industrial site of consumption is likely to require careful review of the existing permits. For some developers, the idea of disturbing what is a settled issue to seek new consents is likely to prove challenging. As with much else in this developing sector, first-of-a-kind projects act as pathfinders in the administrative, consents and regulatory processes. While on the one hand hydrogen is already used in the manufacturing process – so there may be fewer novel technical barriers to consider – the process of obtaining authorisations for new ways of working with low-carbon hydrogen will require new procedures to be developed by the relevant authorities. What's more, as with other pioneering projects, the consent process is likely to require close involvement and consultation with the local community, stakeholders and other interested persons to ensure public awareness and acceptance.

1.5 Contractual structures and revenue considerations

There are currently few public financing mechanisms for the hydrogen industrial market. Instead, many of the available financing options are a mix of public–private funding. The German Federal Ministry of Education and Research is sponsoring the Carbon2Chem project, launched by Thyssenkrupp, by providing over €60 million in funding.[75] The project explores how industrial gases from steel production can be used to create valuable primary products for fuels, plastics or fertilisers, and is expected to render 20 million tonnes of the German steel industry's annual CO_2 emissions economically exploitable in the future, representing 10% of Germany's annual CO_2 emissions produced by industry and manufacturing. In the Czech Republic, there are also examples of public–private collaboration. For example, in the region of Ústí and Labem, Orlen Unipetrol assembled a consortium of 17 public and private entities to sign a memorandum on partnership and cooperation in the development and use of hydrogen as a clean source of energy. The goal of the initiative is to support the use of hydrogen in local industry.[76]

A key legal issue for hydrogen projects in the industrial sector, therefore, is the contractual framework and revenue support mechanisms which may be available to them and which would encourage the switch from grey to low-carbon hydrogen options. Most jurisdictions do not have a prescribed or market-developed set-out of accepted contracts that would apply. Instead, the existing contractual suite from the technology most closely aligned with the relevant hydrogen colour is being adapted for the new uses (for example, International Federation of Consulting Engineers (FIDIC) standard form contracts are being considered for construction contracts involving green hydrogen). While this is an opportunity to create a new market standard and contractual framework that will suit the particular needs of hydrogen projects linked to industrial processes, the current lack of precedents adds to the length and complexity of contractual negotiations.

2. Heating and cooling

2.1 Introduction

With its heavy reliance on fossil gas, heating and cooling is one of the most difficult sectors to decarbonise. This state of affairs is quite pronounced in North America and Europe, where controlling the temperature in buildings is more reliant on gas than electricity. In the United Kingdom, for example, the extensive gas network provides natural gas to over 80% of residential homes and commercial buildings for heating, and heating accounts for over a third of total UK carbon emissions.[77] Nevertheless, a number of countries have identified hydrogen as part of their plan for decarbonising heating, as this sector could be key to achieving net zero aims.[78] Not only is the prospect of hydrogen heating appealing due to its low-carbon production possibilities; it may also be feasible to integrate hydrogen into existing infrastructure without significant overhaul through gas blending or retrofitting the natural gas networks. The most commercially mature and likely technologies for using hydrogen to heat homes and buildings include hydrogen boilers and fuel cells; however, hydrogen may also be used in a hybrid scenario with other heating technologies, such as heat pumps.[79]

A number of innovative projects are being investigated to assess the viability of hydrogen heating. SGN's H100 Fife project in Scotland aims to gain critical insight into hydrogen demand and supply management, security of supply and asset operation in a real-world scenario.[80] The "'world-first' hydrogen heating network" will provide heating to around 300 homes by producing green hydrogen at a dedicated electrolysis plant, powered by a nearby offshore wind turbine, and storing it on site. Similarly, Engie is trialling the injection of hydrogen into the gas distribution network to supply around 200 homes in France.[81] Further south, the Green Hysland project in the Balearic islands aims to create a solar-powered green hydrogen ecosystem, applying hydrogen to various uses including the generation of heat and power for commercial and public buildings.[82]

> **Case study: Singapore data centres**[83]
> Aside from heating, there is a role for low-carbon hydrogen in the cooling of commercial premises, such as data centres, which consume large amounts of energy. With the need for a consistent and reliable energy supply, hydrogen (in particular, hydrogen fuel cells) presents a strong low-carbon alternative to power data centres that could lessen strain on the grid. In Singapore, the

vision for a decarbonised, digital economy has resulted in the advancement of this concept. Keppel Data Centres (KDC), a global data centre owner and operator, has entered into various partnerships to explore new technologies to cool and power data centres. In 2019, KDC teamed up with Singapore LNG Corporation and the National University of Singapore to develop a new energy- and cost-efficient cooling technology for data centres.[84] In 2020, it signed a number of hydrogen-related memoranda of understanding with the following:

- Mitsubishi Heavy Industries Asia Pacific, to jointly explore the implementation of a hydrogen-powered tri-generation plant concept for data centres, producing heat, power and cooling, the aim being to produce hydrogen through a carbon-neutral steam methane reforming process incorporating CCUS technology;[85]
- City Gas and City-OG Gas Energy Services, to explore using liquefied natural gas (LNG) and hydrogen to power KDC's floating data centre park in Singapore, and harness cold energy from hydrogen production for cooling; and
- Royal Vopak, to study the commercial viability of establishing liquefied natural gas and hydrogen infrastructure for power and cooling plants, which may also form part of the floating data centre park project.[86]

As set out in chapter IV, hydrogen is often regulated by domestic gas legislation – in the case of the United Kingdom, by the Gas Act. Given the significant challenge that decarbonising heating presents for the United Kingdom, we consider the legal and other issues here from the UK perspective as they will be common to a number of jurisdictions due to common structures and set-ups.

2.2 Customer-facing issues

As an alternative to gas legislation a number of jurisdictions, including the United Kingdom, are considering developing a regulatory framework for heat networks – distribution systems of insulated pipes carrying heat from a central source (such as heat recovered from industry) to multiple end users.[87] Similar heat networks and district heating schemes operate across continental Europe, providing templates for innovation to permit hydrogen to be used for heating.

Key considerations with heating and cooling are customer experience and social acceptance. Like the idea of driving an electric (hydrogen-fuelled) car, the idea of heating or cooling the home with hydrogen requires socialisation – essentially, public engagement with energy use and management. To this end, there are various pilot projects underway aimed at testing the consumer experience as well as hydrogen's feasibility in residential heating systems. As with any new technology, consumers will likely be eager to learn how hydrogen might work in practice as a heat supply before making any significant commitments or investments. Even where hydrogen is proven to be feasible, consumers may be reluctant to make the switch given the extent of replacement works potentially required.

Compared to electric heating, conversion to hydrogen offers less flexibility in terms of timescales. Wide-scale installations could be postponed until a local or regional authority converts the entire local distribution network, and multiple disruptive home visits may be required. Limited choice in conversion timing could also increase costs if it means replacing functional devices prematurely. And hydrogen-ready alternatives, such as boilers, cooking appliances and fires, are not yet readily commercially available (though a few have started appearing on the market), which may raise questions as to their safety and reliability in a domestic setting.

That said, hydrogen appliances are likely to have similar running costs and performance standards to their natural gas equivalents. The Hy4Heat project mentioned earlier is in part focused on overcoming the challenge of developing new appliances with the aim of delivering prototypes in the early 2020s.[88] Prototype hydrogen boilers are being proactively developed, with leading manufacturers suggesting that commercially available volumes could be ready by 2025.[89] Furthermore, depending on how hydrogen is deployed or used in a heating system, some residential consumers, such as those in flats or apartment blocks, may not notice a difference. A hydrogen boiler installation may require little behavioural change from homeowners, with the heating system operating in a similar manner to its methane gas predecessor.

2.3 Safety and appliance standards

As noted earlier in this Special Report, given the natural characteristics of hydrogen, the legislative regime (largely existing gas legislation) will apply to hydrogen-based heating systems and will impose controls on gas quality and content. There are strict limits on the amount of hydrogen that can be injected safely onto the system. Some uncertainty still exists around the technical and safety testing limitations of hydrogen blending and how this links with the current legislative frameworks. While hydrogen projects may benefit from fitting into well-developed, existing regimes, fresh thinking around the safety case for hydrogen projects in the context of their intended use may be beneficial.

"A hydrogen boiler installation may require little behavioural change from homeowners, with the heating system operating in a similar manner to its methane gas predecessor."

How much hydrogen can be carried in the pipeline?

In the United Kingdom, pursuant to the Gas Safety (Management) Regulations 1996, the concentration of hydrogen that can be safely injected into the UK gas network is 0.1%; demonstration projects such as HyDeploy currently have to undergo higher levels of scrutiny by the HSE in order to be granted an exemption to inject above the current limit.[90] Other countries are also demonstrating hydrogen for heat by blending it into their existing natural gas networks, though still at lower levels. In the Netherlands, physical blending of up to 2% is already achievable with minor adjustments, and the government expects that this could be increased to around 10 to 20%. With more detailed studies, it may even be feasible for the Netherlands' gas grid to handle 100% hydrogen.[91] Portugal's National Strategy for Hydrogen presented the possibility of blending around 22% of hydrogen into the natural gas network without impacting the calorific power of gas in the grid, and as a result, remaining within legislative limits.[92] If research and demonstration projects prove fruitful, current regulations will need to be reviewed to allow for a higher blend.

Consideration will also need to be given to updating the legal requirements for operators undertaking gas work, as well as the standards for hydrogen appliances. And, given the interaction with individual consumers, other legislative considerations will be metering arrangements and the safety testing qualifications of those employed to install and maintain hydrogen-fuelled appliances. For example, only qualified engineers may be able to service a hydrogen boiler and such activities would need to be in compliance with the regulations, codes and guidance used by the recognised industry body for the relevant jurisdiction. In practice this could pose resource constraints while the relevant training is given to ensure compliance, and flexibility may require to be built into the timescales for roll-out of appliances.

Existing gas appliance standards may require modification and updates. In the United Kingdom, these are developed by the British Standards Institute in collaboration with manufacturers, while for international standards the International Standardisation Organisation (ISO) currently has a technical committee responsible for developing standards on systems and devices for the production, storage, transport, measurement and use of hydrogen (ISO/TC 197). In addition, ISO 22734:2019 concerns the industrial, commercial and residential applications of hydrogen generators using water electrolysis, and ISO/TR 15916:2015 concerns basic considerations for the safety of hydrogen systems. Such standards will need to be reviewed and adapted to suit low-carbon hydrogen systems.

2.4 Modification of existing infrastructure

Using hydrogen for heating and cooling is likely to involve modifying existing infrastructure as well as building from scratch. Given the potential retrofitting or repurposing of pipelines and/or the additional uses to which land may be put, such as on-site storage or for hydrogen fuel cells, suppliers and consumers of hydrogen heat should also be aware of relevant land-related issues. These may range from restrictions on how a piece of land, or a building, may be used, to restrictions in lease arrangements on the ability to change existing heating systems to low-carbon, hydrogen-powered systems. While deployment of hydrogen heating could significantly lower a building's emissions and help it to meet energy efficiency and sustainability standards, careful planning and design must be undertaken to ensure this is the case.

2.5 Offtake and price control considerations

In terms of revenue derived from hydrogen heating, there are various considerations (as with the supply of natural gas) in the areas of sourcing, supply and, potentially, control. As with any emerging industry, to obtain sufficient offtake and revenue a growing market will need to be demonstrated.

With the majority of hydrogen heating projects in pilot and testing stages, offtake opportunities are currently limited. However, if blending hydrogen into existing gas networks is shown to be feasible, local demand for hydrogen may increase and gas network operators may therefore be involved in potential offtake arrangements.[93] Demonstration projects are also identifying key users for hydrogen heating. For residential premises, the likely mechanism for supply is through hydrogen boilers, with offtake arrangements working in a comparable way to natural gas boilers; for larger, commercial buildings, a hydrogen fuel cell producing electricity may instead offer a more suitable arrangement.

In terms of attracting private investment, at present most hydrogen projects do not offer enough certainty of demand and fall below the capital requirements of large banks and institutional investors. The majority of hydrogen projects, such as those mentioned in this chapter, therefore rely on government-level financing initiatives or public–private collaborations.[94] For example, the French Government in September 2020 announced a plan for €7.2 billion in public investment to develop a decarbonised hydrogen industry by 2030, with €3.4 billion to be provided by 2023.[95] This is, however, changing, as outlined in chapter VI. With the growing attractiveness of sustainable finance, in terms of both achieving emissions targets and establishing or protecting corporate green credentials, hydrogen projects present a promising opportunity for private sector investors. However, until there is a more stable regulatory framework, an assured coordination of supply and demand and solutions to overcome the current cost hurdles, risk appetite is likely to remain low.

3. Transport

3.1 Introduction

From the expansion of the global electric vehicle (EV) market in recent years it is evident that consumers are willing to adopt alternative fuel sources and are seeking options for decarbonised public and private transport methods. The IEA estimates that while battery technologies are seen as a viable technology for smaller passenger vehicles, partly due to the rapid roll-out seen in recent years, hydrogen-powered vehicles are a complementary option. In particular, hydrogen-fuelled EVs are considered attractive for local public passenger transport, road haulage and commercial vehicles, and for the rail, marine and, possibly, aviation sectors.

Technology readiness levels (TRLs) are at different stages of maturity. Figure 5 shows the TRL of hydrogen fuel cells for the various areas of transport. Those towards the top of the scale are commercially available; those at the bottom require further testing, pilot schemes and development before they will be commercially available.

Figure 5. Technology readiness level of fuel cells for hydrogen-powered transport applications

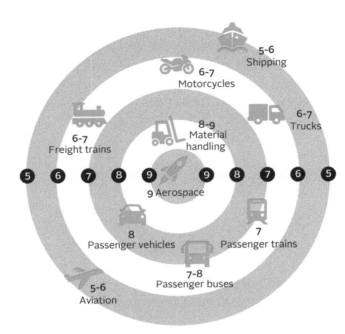

Source: Adapted from "The Norwegian Government's hydrogen strategy".[96]

There are a myriad of regulations that apply in respect of the transport sector, but for the purposes of this chapter we have focused on the legal issues and considerations that apply chiefly to low-carbon hydrogen being used as the fuel for vehicles.

3.2 Road transport

At present, progress on hydrogen-powered transport has been centred around the development of hydrogen fuel cells. In a hydrogen fuel cell a chemical reaction occurs through the combining of hydrogen and oxygen, which generates electricity, heat and water. This is what is used to generate energy. The chemical reaction will continue for as long as there is sufficient fuel (in this case, hydrogen) and therefore offers a potentially longer driving range compared to common lithium-ion battery technologies. Fuel cells are nothing new and are currently used in a number of applications – for example, providing back-up power in facilities such as hospitals, retail and data centres.

While 90% of all global EV sales in 2019 came from China, the United States and Europe combined, almost all sales of hydrogen fuel cell vehicles (FCV) in 2019 came from Japan and South Korea. This momentum has been spearheaded by Honda and, particularly, Toyota, which has been selling its Toyota Mirai passenger FCV on the retail market since 2014. Figure 6 shows the various components within a typical FC(E)V car.

Other reasons that hydrogen FCVs are seen as more attractive options compared to li-ion battery-based EVs centre around speed of refuelling (FCVs typically take around one-third of the time to refuel that an EV battery takes to fully recharge – a high-power rapid charger (600kW) can charge an EV battery for a range of 400km in around 60 minutes, while a hydrogen refuelling pump can fill an empty FCV tank in under 20 minutes)[97] and higher capacity density in terms of how much energy/power it can provide. This means that FCVs are more appropriate than EVs as larger road vehicles like SUVs and heavy-duty vehicles or fleet vehicles. However, a key component to their success will be having the necessary refuelling infrastructure.

A gateway into the hydrogen mobility sector appears to be through the uptake of hydrogen-fuelled public transport. The United Kingdom has already seen the launch of zero-emission hydrogen buses, which were partly funded through Transport for London and the European Commission's Fuel Cells and Hydrogen Joint Undertaking (FCH JU),

Figure 6. Fuel cell technology of a hydrogen FC(E)V explained

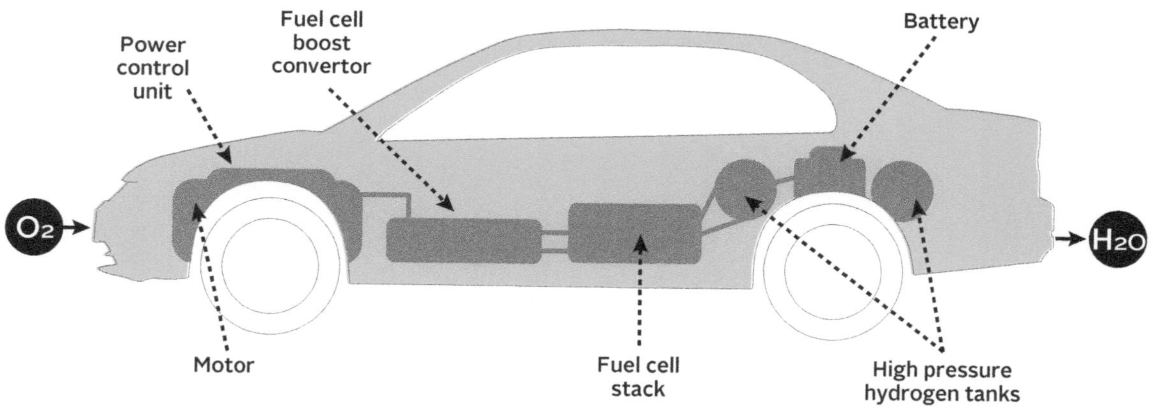

Source: Author.

"Roll-out of the necessary infrastructure to support hydrogen-fuelled vehicles is a chicken and egg scenario: where the number of hydrogen vehicles increases, so too must the number of hydrogen fuelling points; but equally, access to fuelling points will boost the number of hydrogen vehicles being purchased."

and has committed to 3,000 buses by 2024.[98] Spain, too, is in the
process of rolling out a fleet of 5–10 hydrogen-fuelled buses and a
fuelling station as part of the Green Hysland project in Mallorca,
as well as eight fuel cell buses in Barcelona. Similar plans are also
being considered for the capital, Madrid, after the success of the
HyFleet:Cute pilot project, which ended in 2009 and saw hydrogen-
fuelled buses integrate into the public transport system. Further, as a
result of German government transport initiatives, Regionalverkehr
Köln GmbH (RVK) now operates the largest fleet of fuel cell buses in
Europe to date: 35 buses, increasing to 50 by the end of 2021.

This is possible, in part, because in many countries the legislation is
either silent on the use of hydrogen FCV-type vehicles, or permits
their use as part of the wider legislation encouraging use of biofuels.
That said, vehicles powered by hydrogen will still need to comply
with local vehicle safety regulations, which in the case of EU member
states are set out in, amongst others, Regulation (EC) 661/2009
(governing the type-approval requirements for the general safety
of motor vehicles, their trailers and systems, components and
separate technical units) and Regulation (EU) 79/2009 (setting
out requirements for the type-approval of motor vehicles with
regard to hydrogen propulsion, including hydrogen components and
hydrogen systems and the installation of such components and
systems). Although there are few hydrogen FCVs on the roads at
present, obtaining approvals for their use on roads is not reported
to be a major legal impediment, even given the lack of experience
in hydrogen vehicles at local authority level. However, approval needs
to go hand in hand with a clear and unified set of rules for service
and inspections. While there are overarching rules (such as the EU
roadworthiness package) about how periodic roadworthiness tests,
vehicle registration and inspections are carried out, putting these
requirements into practice and having the relevant trained resources
available to carry out the various tasks is not yet a reality in most
jurisdictions.

3.3 Fuelling with hydrogen

In some countries – and in most of the European Union as a result of
the Renewable Energy Directive 2009 – hydrogen is recognised as a
transport fuel. Where this is the case, national laws tax hydrogen as
a fuel, apply emissions reporting standards against it and have
standards for fuel quality that hydrogen must meet. Further, because it
is recognised as a low-carbon transport fuel, in some cases low-carbon
hydrogen (chiefly, green hydrogen) may be eligible for certificates of
origin (but see below with respect to traceability issues).

However, in many countries, including the United Kingdom, there is
currently no specific standard that binds hydrogen fuel suppliers with

regard to the fuel quality that must be supplied. Instead, this will be a matter of contractual agreement between the producer and the fuel supplier. This may be achieved by reference to international standards that apply for hydrogen use as an industrial gas (in particular ISO 19880-8, which specifies the quality of the gaseous hydrogen at hydrogen distribution facilities and hydrogen fuelling stations for proton-exchange membrane (PEM) fuel cells for road vehicles).

As noted, key barriers for hydrogen are the lack of infrastructure at most forecourts, and the prevailing regulatory treatment. Roll-out of the necessary infrastructure to support hydrogen-fuelled vehicles is a chicken and egg scenario: where the number of hydrogen vehicles increases, so too must the number of hydrogen fuelling points; but equally, access to fuelling points will boost the number of hydrogen vehicles being purchased.

Case study: South Korea's hydrogen cars[99]

In April 2021, the Korea Automobile Manufacturers Association (KAMA) announced that South Korea had the highest car distribution rate in the world for hydrogen vehicles.[100] Globally, South Korea hosts 33% of operating hydrogen cars, and the numbers continue to rise. According to KAMA, from 2016 to 2020 the average annual growth rate of hydrogen cars was 235%, compared to 2% for internal combustion engines, and 88% for electric cars. This reflects the ambition of the South Korean government, outlined in its 2019 Hydrogen Economy Roadmap, to have 5.9 million fuel cell passenger vehicles on the road by 2040.[101]

However, to keep pace with the high – and growing – vehicle numbers, South Korea's hydrogen charging infrastructure requires expansion. The ratio of vehicles to charging point stands at 180:1, while in other nations such as Germany, the ratio is 9:1. Japan and China have ratios of 38:1 and 56:1 respectively. It is far from certain that South Korea's charging points will develop at a rate fast enough to keep up with its rapid deployment of hydrogen vehicles, and KAMA recognises that there is an issue. Some provinces have no charging stations at all. With a view to developing the infrastructure, in 2019 the South Korean Ministry of Trade, Industry and Energy established a special purpose vehicle (SPV), HyNet, in collaboration with 13 leading South Korean industrial companies. The aim is to build 100 hydrogen charging stations by 2022 – around a third of the government's stated target of 310.[102]

In most cases, it is unlikely that hydrogen for fuelling vehicles would be produced at the site (see discussion in chapter IV of the legal issues regarding transportation and long-term storage of hydrogen). At the depot level, hydrogen storage in cylinders throws up its own challenges – not least that changes to the infrastructure will likely require fresh authorisations. While many countries have consenting regimes relating to large-scale production of industrial gases (such as hydrogen), the low-volume distribution and storage rules may not yet exist.

Despite the first hydrogen refuelling station having been installed over ten years ago, most countries still do not have specific legislation for hydrogen refuelling. Nevertheless, many countries believe public transport to be the key area for using hydrogen in the mobility sector. So, operators of hydrogen refuelling stations will need to navigate the wider legal and administrative rules around storage of chemicals and flammable and dangerous gases. For example, depending on the duration of storage and the quantities involved, consent may be required under local hazardous substances regimes (such as the Control Of Major Accident Hazards Regulations 2015 in the United Kingdom). This would be in addition to navigating the complexity of the ADR framework around the transportation of hydrogen by road – something that is likely to apply at least for the last one-mile segment of hydrogen transportation.[103]

3.4 National and regional initiatives to promote hydrogen-powered vehicles

Initiatives to promote the use of hydrogen-powered vehicles are becoming more common, most notably in Europe where, in March 2020, 44 major companies from across the transportation supply chain signed a letter of intent headed, "Joint call for the deployment of hydrogen fuel cell trucks: a needed shift towards a carbon-neutral society".[104] The signatories pledged to deliver between 5,000 and 10,000 hydrogen-powered trucks and establish at least 100 hydrogen refuelling stations across Europe by 2025. This pledge was strengthened in November 2020 when a statement was delivered to the European Commission by 62 companies committing to 100,000 fuel cell heavy goods vehicles (HGVs) and 1,500 refuelling stations by 2030.[105] Among the signatories were original equipment manufacturers (OEMs) (Honda, Daimler, Hyundai), equipment manufacturers/OEMs (Ballard, Michelin, Toyota), infrastructure and hydrogen providers (Engie, Air Liquide, Linde), truck operators and logistics users (Heineken, Unilever, FM Logistics) and other associations (Port of Rotterdam, Waterstofnet).

The EU has often been at the forefront of subsidy and investment in the area of new energy technologies, and hydrogen is no exception. The European Commission has been responsible for a number of initiatives to advance hydrogen technologies in the transportation sector. The Joint Initiative for Hydrogen Vehicles across Europe (JIVE), begun in 2017, has the aim of deploying around 140 zero-emission fuel cell buses, as well as the necessary refuelling infrastructure, across five EU countries. JIVE is co-financed by the FCH JU, with a grant of around €32 million, and is supported by a consortium of 22 partners. An extension of the initiative, JIVE 2, was launched in 2018 to deploy around an additional 150 fuel cell buses within major European cities. In November 2020 it was announced that over 200 hydrogen fuel cell buses have been ordered through the initiative, with 50 buses already in operation across three cities: Cologne, Wuppertal and Pau.[106] JIVE and JIVE 2 aim to commercialise the deployment of fuel cell buses and consequently unlock economies of scale, driving down costs of manufacture and maintenance.

Additionally, the European Commission is funding two research projects that seek to deploy an additional 1,400 hydrogen-powered cars, vans and trucks, as well as 49 hydrogen refuelling stations across the EU by 2022. Hydrogen Mobility Europe (H2ME 1) and Hydrogen Mobility Europe 2 (H2ME 2)[107] will receive a total of around €67 million in funding under the EU's Horizon 2020 initiative, and have support from over 40 partners across the industry, including BMW, the University of Manchester and Rijkswaterstaat (the Dutch Ministry of Infrastructure and Environment).

3.5 Rail transport

In 2016 Alstom, with support from the German Government's National Innovation Program for Hydrogen and Fuel Cell Technology (NIP), developed the world's first passenger train powered by a hydrogen fuel cell – the Coradia iLint.[108] Since then, Alstom have rolled out hydrogen-powered trains in Germany, Austria and the Netherlands, where trials of the Coradia iLint alongside regular service trains have been successful. Alstom have also signed deals in France and Italy and are working closely with governments across the Middle East, where there is appetite for making public rail transport more environmentally friendly.

In the United Kingdom, engineers from the University of Birmingham and Porterbrook have worked together to produce the United Kingdom's first hydrogen-powered train, HydroFLEX, which completed successful mainline tests in September 2020.[109]

Many of the issues noted above for the road transport sector with regards to fuelling apply also to the rail sector.

3.6 Marine transport

To date, international shipping and aviation have been largely exempt from the changes in legislation aimed at driving down carbon emissions. While some efforts have been made around carbon emissions trading and, recently, the sulphur content in shipping fuel, the opportunities for decarbonising jet fuel or shipping fuels are only just being recognised.

A key driver is the United Nations International Maritime Organization (IMO) ambition to reduce GHG emissions from international shipping by 50% compared to 2008 levels by 2050.[110] Given that the IMO's remit includes oversight of key international conventions (which are often then imported into national laws) such as the International Convention for the Safety of Life at Sea (SOLAS), the International Convention for the Prevention of Pollution from Ships (MARPOL), and codes such as the International Code of Safety for Ships Using Gases or Other Low-Flashpoint Fuels (IGF Code), it stands as a key organisation for promoting and regulating the uptake of hydrogen or ammonia as a decarbonised fuel in the maritime sector. That said, while some discussion has been taking place around the use of hydrogen, much of it is in the early stages.

A key consideration for shipping is whether to select liquid hydrogen or synthetic ammonia as the fuel type of choice. Currently, creating green hydrogen is energy-intensive, requires significant storage space and is not very efficient (the end-to-end process is reported to have an efficiency of about 40%, though improving). Burning ammonia made from green hydrogen may be a more attractive alternative. Ammonia can store almost twice as much energy as liquid hydrogen by weight and has nine times the energy density of lithium-ion batteries. Although it is mainly used as an industrial fertiliser, its potential as a transport fuel has been demonstrated by NASA in its deployment in rockets.

While many of the transport and storage issues identified in this chapter apply to the maritime sector, a key challenge peculiar to shipping is cost. At current prices, decarbonising the shipping sector using ammonia would require $1–1.4 trillion of capital investment from 2030 to 2050, according to the Global Maritime Forum.[111]

What's more, national laws and regulations would need to be developed around standards that would apply to ammonia-fuelled vessels. Some companies, like Shell,[112] are planning to move into this sector and certain countries are starting to develop these rules. For example, as part of Norway's Hydrogen Strategy, it plans to develop regulations and standards for the marine and shipping industry to ensure that the Norwegian Maritime Authority and Norwegian Coastal Administration have the knowledge to handle new green shipping solutions, including an understanding of hydrogen regulation.[113] Such developments highlight the important role hydrogen will play in the future of the maritime sector.

3.7 Aviation

In the case of the aviation sector, hydrogen is an attractive replacement for kerosene – in its liquid form it is light enough to be carried and when burnt it produces water instead of CO_2. Alternatively, aircraft may have electric motors powered by hydrogen fuel cells (similar to those being considered for road and rail transport).

Rules for using hydrogen to fuel aircraft are effectively absent. A robust legal framework for transporting, storing and refuelling with liquid hydrogen is critical in order to accelerate deployment of hydrogen in aviation. Without this, the relevant regulator (such as the European Union Aviation Safety Agency (EASA)), which must authorise the aircraft (including its parts) before it is permitted to fly, is unlikely to be able to grant consent. Given the high risk in event of aircraft failure, the regulator must ensure not only that the individual components work, but that they interact safely with each other and with other technical elements of the aircraft.

Therefore, while major aircraft engine manufacturers are considering developing hydrogen-propelled aircraft, seeing one in the skies is some years or even decades away. A McKinsey study commissioned by Clean Sky 2 JU and Fuel Cells and Hydrogen 2 JU, for example, reports that reliable hydrogen-fuelled aircraft would be feasible by 2035.[114]

At present, efforts in the aviation sector appear to be largely directed at information gathering and research and development. For example, in February 2021, the European Commission announced its proposal for a future European Partnership on clean aviation under the Horizon Europe programme for research and innovation, which will pursue technologies to enable hydrogen-powered aircraft.[115]

3.8 Legal challenges and opportunities

When it comes to legal issues, common themes apply to using hydrogen in the transport sector generally.

A number of the projects in the transport sector are being developed with public authorities as customers, and often as funders. While funding is key to enabling the roll-out of hydrogen-fuelled vehicles and the relevant infrastructure, with funding come questions around eligibility, stability and sufficiency, as well as issues of state aid and procurement.

Given that contracting for public transport and associated infrastructure often involves a government authority as the counterparty (whether for associated land rights, leases of the transport or other matters) hydrogen buses could be the catalyst for the build-up of the administrative knowledge and processes. Assuming regulation and funding can be implemented purposefully and

"Assuming regulation and funding can be implemented purposefully and effectively, hydrogen-fuelled transport in general has the potential to mirror the rapid uptake of EVs seen in some countries."

effectively, hydrogen-fuelled transport in general has the potential to mirror the rapid uptake of EVs seen in some countries.

Together with funding, the ability to assess whether hydrogen is low carbon (especially when it is co-mingled with non-low-carbon hydrogen) will be needed from both environmental accounting and subsidy support perspectives. For example, in a number of European countries hydrogen is defined as a renewable transport fuel and so is eligible for certificates of origin. However, at present this system of certification is not harmonised. The absence of a common definition of 'low-carbon hydrogen', for example, is a barrier to choosing hydrogen as an alternative fuel, and may well also be a barrier to the ability to transfer hydrogen between different regulatory systems.

For the purpose of traceability, hydrogen fuel dispensed into vehicles (or transferred to another customer) must be adequately measured and tracked to allow the end consumer to know the precise amount of hydrogen it consumes. This is necessary for:

- pricing;
- accounting for the amount of green hydrogen dispensed to which certificates of origin may apply; and
- taxes on hydrogen as a fuel.

Further, while some countries (such as the United Kingdom) offer tradeable certificates for green hydrogen use, the complexity of tracing hydrogen from source of production to end user creates interface risks in the contractual structures which further impede the development of a green hydrogen market.

Another common challenge in the sector is matching supply with demand. Hydrogen producers are reluctant to produce hydrogen for fuel without knowing there is a ready demand from transport users, while vehicle manufacturers are reluctant to invest in bringing new FCV vehicles to the market without assurances that such vehicles will have access to necessary refuelling infrastructure and fuel. Without a sufficient supply of low-carbon hydrogen to power the relevant vehicles on a regular basis, these uncertainties introduce levels of risk which none of the contracting parties is willing to bear. As discussed in chapter VI, the ability to raise equity or debt for the capital investment required is equally challenged by this unpredictability.

Growth in production is expected to address the other common (commercial) barrier currently faced by green hydrogen – that its production costs are currently at least double those of grey hydrogen and, in the case of aviation, at least six times higher than the production costs of kerosene.

Finally, in the case of aviation, although low-carbon hydrogen does not produce CO_2, at high altitudes it is known to produce nitrogen oxides as well as water vapour. The precise effects of this on the aims of addressing the adverse impacts of climate change are not fully known at present.

Case study: Wrightbus

Wrightbus is a Northern Ireland-based manufacturer of zero-emission hydrogen buses. The owner, Jo Bamford, also owner of hydrogen firm Ryze Hydrogen, has outlined the company's intention to introduce 3,000 hydrogen buses to the United Kingdom by 2024.[116] While there have been a number of bus operators expanding their fleets of electric buses, due to current battery ranges they are often not suitable for all routes. Hydrogen fuel cells can be a more suitable alternative for heavy vehicles, particularly for buses, as refuelling infrastructure can be installed at central depots. In 2020, Wrightbus launched the world's first fleet of hydrogen double-decker buses. In March 2021, Ballard Power Systems received an order from Wrightbus for 50 fuel cell modules to power the vehicles, signalling the deployment of 50 new hydrogen-powered buses in a number of UK cities, including Birmingham, Aberdeen, London and Belfast. The buses are to be partially funded under the JIVE programme.[117]

Also in March 2021, the Department for Business, Energy and Industrial Strategy (BEIS) announced an £11.2 million funding package to develop and manufacture low-cost hydrogen fuel cell technology for buses and to create a hydrogen centre of excellence with Wrightbus in Ballymena, Northern Ireland.[118] The funding is being coordinated by the Advance Propulsion Centre, which supports the development of low-carbon-emission technologies for cars, buses, heavy goods vehicles and vans. BEIS anticipates that investment in innovative technologies such as hydrogen fuel cells will not only support a green recovery from the impacts of the COVID-19 pandemic, but also position the UK as a global leader in automotive technology.[119]

Case study: refuelling stations and fuel cell vehicles in Japan[120]
Japan has one of the largest hydrogen refuelling networks in the world. According to the Fuel Cell Commercialisation Conference of Japan, as of 21 November 2020 Japan had built 146 commercial hydrogen refuelling stations, of which 106 were fixed and 40 were skid-mounted.[121] Japan has a clear ambition to develop this network further, setting a target to have 900 hydrogen refuelling stations by 2030 in its third Strategic Roadmap for Hydrogen and Fuel Cells.[122]

These numbers reflect Japan's overall commitment to developing a hydrogen (vehicle) economy, which is not new. In the 2000s, government and industry in Japan focused on promoting FCVs and refuelling stations. In 2008 the Ministry of Economy, Trade and Industry published a plan aiming to deregulate both FCVs and refuelling stations, as well as promote technical innovation.[123] For the 2020 financial year, the Japanese government budgeted ¥70 billion to support the promotion of hydrogen, of which ¥30 billion was allocated to subsidise hydrogen fuel cell vehicles and fuel station construction.[124]

While progress regarding refuelling infrastructure has been impressive, the Japanese hydrogen fuel cell vehicle market is yet to gain as much traction. Car manufacturer Toyota launched the Mirai FCV in December 2014 but the high sale price has been an inhibiting factor. To encourage mass appeal, Toyota aims to reduce the price of fuel cell vehicles to around $20,000 by 2025.[125] The government also plans to achieve a cost reduction of FCVs to the level of hybrid vehicles by around the same time.[126]

The only country to achieve its FCV sales targets in 2020 was China. However, Chinese FCV strategy focuses on commercial vehicles such as hydrogen-powered heavy-duty trucks, in contrast to the passenger vehicle focus of Japan, thus eliminating much of the consumer choice and cost element in making the transition.

However, Japan is also exploring fuel cell vehicles beyond the retail market. In January 2020 Honda and Isuzu agreed to collaborate to research fuel cell trucks,[127] and in October 2020 the East Japan Railway company, JR East, agreed a business partnership with Toyota and Hitachi to develop and test railway vehicles that use hydrogen-powered fuel cells and storage batteries as their electricity source.[128] Testing is scheduled to start in March 2022.

4. Energy system management

4.1 Introduction

Management of the electricity grid has always been a careful balancing act, usually overseen and managed by the system operator working in close cooperation with other network operators. However, as the electricity generation profile becomes harder to forecast and control, with increased amounts of intermittent generation exporting to the grid, both the system operator and the intermittent generators are seeking ways to optimise and store the electricity generated. For example, in the United Kingdom renewable energy generation is expected to exceed demand in 48% of all hours by 2035, which would mean that 20% of that renewable energy generated could go to waste if there is insufficient demand or be constrained if the energy cannot be exported to the grid.[129] This is where hydrogen is seen as a potential tool, especially for longer-term (over four hours but especially seasonal) storage.

Using hydrogen for storage would allow renewable energy to be stored and prevent generation stations being turned off (or constrained) at times of surplus. Currently the management of these instances is balanced by the system operator using coal- and gas-fired power plants, which, without carbon capture, take away from the decarbonisation efforts.

This is not dissimilar to the role that gas storage has played historically in facilitating the efficient functioning of the energy market and providing an important source of flexibility that is needed to address differences in non-storage sources of supply and demand. Energy storage management involves different storage needs ranging from strategic (to be used in emergencies) to seasonal (where hydrogen would be injected in the summer and used in the winter).

Typically, gas storage facilities are located underground in depleted gas reservoirs or salt caverns. The storage facilities located above ground are usually at low pressure and with smaller capacities, though this may be different in some jurisdictions and especially for LNG facilities.

Stakeholders in the use of hydrogen for system management broadly fall into the following categories:

- system operators;
- storage operators;
- storage customers; and
- energy use customers.

Each of these have their own drivers and interests in how they would see hydrogen serving their business needs.

While the system operator is interested in the overall management of the energy system, the costs of building and operating the store may not be recoverable under its licence through tariffs charged to consumers. Rather, its requirements for addressing system balancing needs and system optimisation services may be limited to those procured where hydrogen is sold as a service. From a system operator's perspective, the value from hydrogen would be a more efficient functioning of the overall energy system due to an increase in the transmission load factor, meaning hydrogen would play a role akin to other system services procured by the system operator (such as black start, reactive power and other ancillary services).

"Management of the electricity grid has always been a careful balancing act, usually overseen and managed by the system operator working in close cooperation with other network operators."

Operationally, at least at present, a number of system and network operators may also be prohibited by their licence conditions from injecting hydrogen into the existing gas grid infrastructure. For example, in the United Kingdom, the Gas Safety (Management) Regulations 1996 limit the amount of non-methane gas transported on the gas grid, and changes to this would require the safety case to be approved by the HSE as part of overall system and safety management considerations, as well as with a view to minimising the risk of supply emergencies.

There is an additional dimension to consider in the many jurisdictions where there has been some divergence on how the gas and electricity grids operate, even if there may be an overall single gas and electricity system management function (eg, with the regulator). This has resulted in the development of market standards and practice in gas distinct from those in the electricity sector (and vice versa). Using hydrogen for energy system management has the potential to bring these complementary sectors closer together again, in tandem with more short-term energy management tools such as use of batteries.

In the case of the storage operator, the ability to offer hydrogen as a service will be governed by the licence conditions and the commercial

"A storage operator may be able to derive additional revenue by offering the capacity to 'park' hydrogen when the price is lower and sell on when the price is higher (exploiting either a seasonal or other forecast spread)."

agreements it is able to enter into with its customers. For example, a storage operator may be able to derive additional revenue by offering the capacity to 'park' hydrogen when the price is lower and sell on when the price is higher (exploiting either a seasonal or other forecast spread). Plus, using a hydrogen store may help manage commercial risk in take or pay contracts and optimise hydrogen deliveries from multiple sources.

However, this assumes that the storage operator has an asset ready to take the hydrogen to store. Currently this is a key commercial barrier to the development of hydrogen storage facilities – the multi-million capital expenditure needs to be justified and costed against future demand and associated revenues. Accordingly, in addition to the legal hurdles of needing to create a suite of contractual documents suited to hydrogen trading activities, storage operators would at present face significant commercial challenges to raising the financing needed for the construction of the facilities.

Hydrogen storage in liberalised markets

Historically, energy storage (gas storage) facilities were built and operated by a centralised system operator with overall management responsibility for the energy system. This allowed the system operator to include the significant costs of building a store within its tariffs.

However, liberalisation of the electricity and gas markets has resulted in the separation of activities, and restrictions on the types of activities that the system operator may carry out. In the European context, the EU Third Internal Energy Market Package, in force since 2011, introduced a number of measures to promote greater transparency and competition in the market. In particular, transmission system operators and storage facility owners are required to provide non-discriminatory access to their facilities and give customers (in this context, the user of the hydrogen store) sufficient information about the services they provide and how to access the system. Further, in setting their menu of services, storage operators are required to offer:

- long- and short-term services;
- firm and interruptible services; and
- bundled and unbundled services.

All of the measures were drawn up with a methane (and LNG) gas network in mind and in response to a historical context of expanding the range of entities able to gain access to (in many cases) an existing set of assets.

The way the European unbundling requirements have been implemented at national levels may also be relevant. In a number of jurisdictions, for example, the gas laws do not apply to gas storage offshore, and there is therefore some ambiguity regarding whether exemptions or clarification are necessary for the storage of hydrogen offshore.

Where unbundling laws apply, a number of legal challenges present themselves for hydrogen storage operators. If they are to be legally unbundled from relevant affiliates, and their facility to operate independently assured, they will bear the burden of raising significant capital against service and usage contracts with third parties for hydrogen production and storage at the relevant facility. What's more, in certain circumstances the storage operator may also be required to provide a parent company undertaking with regards to its financial planning and levels of indebtedness. While standard in many other parts of the energy sector, this should also be factored in when planning to operate a hydrogen storage facility.

Conversely, a liberalised market structure also presents an opportunity. By creating a framework for a new secondary market in hydrogen trading, similar to methane, a storage operator is incentivised by its legal requirements and runs auctions for capacity allocation as part of ensuring appropriate economic signals for the efficient and maximum use of the store's capacity. In due course, as the market evolves, the allocation mechanism can be flexed to respond to evolving market circumstances.

Hydrogen for energy management involves producing hydrogen using – in the case of green hydrogen – renewable electricity at times of excess capacity on the system. This hydrogen can then be stored and converted for use in power system applications or used in hydrogen fuel cells at times when there is less capacity on the system. Salt caverns, pressurised tanks and repurposed gas pipelines are currently expected to be the most likely short- to medium-term hydrogen storage options.

From the perspective of system management, hydrogen can therefore fulfil the role that other baseload technologies like methane gas have fulfilled in the past. It can also complement other shorter-term storage options such as lithium-ion batteries.

Case study: using hydrogen to manage the Danish power grid

Air Liquide's HyBalance project produces hydrogen using a 1.2MW electrolyser in Hobro, Denmark.[130] Operational since 2018, it is one of Europe's first facilities to produce hydrogen by PEM water electrolysis on an industrial scale. The plant has been used to demonstrate how hydrogen can accommodate the intermittency of renewable energy generation and so contribute to the management of the Danish electricity system. It also produces hydrogen for transport via a pipeline and onward sale to an industrial customer. With 120 tonnes of hydrogen delivered since 2018, it has also provided hydrogen for fuelling stations in Copenhagen, fuelling a fleet of hydrogen fuel cell-powered taxis. The project is led by Air Liquide, with Cummins, Centrica Energy, LBST and Hydrogen Valley and was funded by FCHJU (€8 million) and Danish Energy Technology Development and Demonstration Programme (EUDP) (€2.6 million).

4.2 Traceability of low-carbon hydrogen

As discussed in the context of decarbonised transport, end users will wish to ensure that hydrogen can be traced from production source through to consumption. This is both a practical question – how to trace that a molecule of hydrogen injected for transportation onto a network system and stored in a storage facility is still ultimately the equivalent low-carbon hydrogen molecule by the time it is consumed – and a legal one.

A common theme in the barriers faced by those seeking to produce and consume low-carbon hydrogen is the lack of an integrated, joined-up legal framework, and the issue comes up again in the context of energy system management.

It is anticipated that energy storage will be encouraged in some legal frameworks. For example, the European Union's Renewable Energy Directive anticipated supporting the use of energy storage systems to facilitate the integration of intermittent generation from renewable sources into the overall electricity system, and provided for hydrogen to fit the definition of 'renewable gas' to benefit from the Guarantees of Origin scheme.[131] And the 2019 version of the European Union's Electricity Directive strengthened the ability to use hydrogen for energy storage – the definition of 'energy storage' anticipated "the conversion of electrical energy into a form of energy which can be stored, the storing of such energy, and the subsequent reconversion of such energy into electrical energy or use as another energy carrier".[132] However, when hydrogen is stored in its liquid form, the interplay with gas legislation adds unintended consequences with regard to hydrogen injected into, and transported through, the gas system, which will need careful managing.

"It is difficult to envisage a world, without significant changes in the market, where ordinary consumers of electricity or gas would be interested in analysing the myriad of legal provisions applying, or setting themselves up as energy storage systems for the purpose of low-carbon hydrogen storage."

It is difficult to envisage a world, without significant changes in the market, where ordinary consumers of electricity or gas would be interested in analysing the myriad of legal provisions applying, or setting themselves up as energy storage systems for the purpose of low-carbon hydrogen storage. Nonetheless, end users such as corporate entities looking to demonstrate their decarbonisation credentials would wish for the incentive of a tradeable and traceable instrument like a guarantee of origin certificate to demonstrate that their energy use is low carbon. The absence of this traceability puts hydrogen, and especially hydrogen which passes through multiple interfaces and is therefore harder to net off against a production source, at a disadvantage compared to other renewable sources of generation.

An option for renewable electricity generators to own and operate hydrogen storage facilities is self-evidently attractive. They can choose how to deal with the electricity generated and arbitrage export to the grid versus use for storage based on market signals, including signals from the system operator looking to reduce constraint payments or seeking additional back-up sources of electricity. However, operation of a storage facility may be incompatible with unbundling the rules operating in many liberalised energy markets. For example, in the European Union, gas producers are not permitted to run and operate a gas storage facility save in very discrete circumstances, but the position is not as clear for electricity generation.

Case study: transit barriers

Electricity and gas interconnectors have evolved to provide system security as well as cost benefits across the regions they interconnect. Pool systems allow national boundaries to be bypassed in favour of energy system management efficiencies that benefit a wider set of consumers. This is the result of many years of legislative and technical effort across a network of operators and supportive governments. Unfortunately, these are not systems that hydrogen projects can readily plug into at present, because the parameters for gases transported on national gas systems are not always as well harmonised and integrated as those for electricity interconnection. As a result, in regions like the European Union, what is an acceptable level of hydrogen to transport in Germany (10 mol%) is significantly higher than what is permitted in the Netherlands (0.02 mol%).[133] This trade barrier would mean that countries with different national standards could reject hydrogen coming into their jurisdictions. As was achieved for methane gas, setting common standards and uniform network rules by way of agreements or network codes, as between system operators, would enable adjacent systems to allow interconnection around a given interconnection node.

VI. M&A and financing issues in hydrogen projects

1. Introduction

While use of hydrogen in existing applications (especially industrial processes) is not new, the development of low-carbon hydrogen options is creating new technology classes and therefore attracting new investors to this area. Traditionally, the large industrial players (for example, Linde plc and Air Liquide) or steel manufacturers (such as Tata Steel and ThyssenKrupp) produced hydrogen on site for their manufacturing processes. As such, hydrogen production and the nearby on-site demand was (and largely remains) an integral part of manufacturing, with limited outside investor involvement.

As with any developing technology, the scale of investment activity is currently small. Indeed, existing projects are largely publicly sponsored, with limited private sector participation. This is changing. More partnerships and joint ventures are beginning to emerge, with additional players seeking to find attractive investment cases. For example, Hyundai and INEOS signed a memorandum of understanding in late 2020 to jointly undertake investigations into new opportunities within the hydrogen economy, notably working together on the development of INEOS' first fuel cell vehicle, Grenadier.[134] This partnership signifies INEOS' diversification into FCVs and the low-carbon hydrogen market, while giving Hyundai's fuel cell technology an edge in the European market. Similarly, the NortH2 project is backed by a consortium of Shell, Gasunie and Groningen Seaports and, as of December 2020, was further strengthened by Equinor and RWE joining as partners.[135] Currently Europe's largest green hydrogen project, NortH2 aims to have an integrated offshore wind to green hydrogen capacity of over 10GW by 2040. Being backed by a consortium of key renewables players means that NortH2 is able to develop on a large scale which, it predicts, could lead to a 20% reduction in societal costs when compared with a small-scale approach.

2. Key structures

For green hydrogen projects, where the focus is on structuring the project around an electrolyser, two key considerations are:

- assessing where the electricity to power the electrolyser will come from; and
- ensuring that the hydrogen produced has a market to which it can be sold.

Typically, the electrolyser manufacturer does not have the skills necessary to develop and construct the infrastructure associated with hydrogen production so may need to partner with another entity that can assist with these aspects. Depending on the interests of the parties involved, the project may be driven from the electricity generation end, where the electricity producer sees the benefit of adding an electrolyser to an existing project (or one in planning stage), so would tie the electricity supply to the electrolyser from the outset; or, from the hydrogen offtake end, where there is a need to meet a preset demand (such as volumes of hydrogen needed in a transport scheme or as part of a manufacturing process). In the latter case, the delivery of a steady supply of quality hydrogen tends to influence how the other contractual documents fit around the demand.

In the case of hydrogen produced by other means (for example, blue hydrogen), the structure would also need to take into account the infrastructure for the capture, transport and storage of CO_2 produced by the project. Often the entities involved in CCS processes would be separate from those involved in the production of hydrogen, so a key part of the structuring of the overall project will require contracting the interface arrangements as between these processes.

Finally, given that many existing hydrogen projects currently involve a degree of public authority involvement, the role that a state authority may wish to have in the development and operation of a hydrogen project will be a key consideration. For example, an authority may assist with obtaining certain permits or may impose conditions on the project as part of its state funding.

3. Key terms and requirements for investment

As many hydrogen projects are still in early demonstration and development stages, there is limited M&A and financing experience in the market, and no typical contractual suite that will apply.

That said, undertaking development of new technologies beyond pilot or R&D stages is often too risky and complex for a single company, and a number of private sector joint ventures, often with public support, have emerged to share the risks appropriately and pool the knowledge of the companies involved. This is to be expected, since M&A activity often increases after a project has progressed sufficiently from the development window (ie, once the relevant project has secured the appropriate planning, land and electricity connection and hydrogen export rights). As is seen in the wider renewables market, at this point the capital needs of the project necessitate a share or asset sale to bring in more capital.

4. Project company-based joint venture approach

A key requirement for investment from the sponsors is likely to be the use of an SPV that holds the project rights and assets for a single hydrogen project. Such a structure will inevitably evolve according to the requirements of the funding arrangements (eg, portfolio, multiple

"As many hydrogen projects are still in early demonstration and development stages, there is limited M&A and financing experience in the market, and no typical contractual suite that will apply."

debt structures and cross-collateralised structures) and of a more varied sponsor group as the technology develops.

Some key terms and requirements of sponsors of hydrogen projects will therefore be based on this evolutionary experience, as well as the lessons learnt from the successful development of other new technologies, notably in the renewables sector.

Sponsors embarking on a joint venture to develop a hydrogen project will be keen to understand the scope of their involvement. This may be influenced by the wider aims of the project, such as to break into a new market, gain knowledge or beat the competition. The terms of the sponsor arrangements will need to reflect how the sponsors will try to achieve these desired outcomes and set some parameters around the criteria by which the project's success will be judged. Typically the terms of agreement between the sponsors will need to set out clear approval requirements, with enhanced controls on reserved matters (which often require more scrutiny) and the arrangement as to how the project will be managed. Equally important, the terms need to enable a sponsor to exit in certain scenarios, with a process in place to ensure that the overall stability of the project is not adversely impacted. This will include consideration of rights to use intellectual property, assessment of the funding positions, and compliance with any agreed restrictions or regulations. While we have not seen many sales to third parties or any initial public offerings (IPOs) of hydrogen projects, in due course planning for such events will need to be part of the M&A considerations.

In addition, prior to embarking on the joint venture or acquisition, it is important to assess the regulatory and competition impacts in order to avoid surprises at later stages of project development. These include:

- being aware of any requirements for additional compliance with competition regulations;
- ensuring that the relevant permits and authorisations are situated with the right company; and
- understanding the implications from a tax and state aid perspective.

In addition, merger control requirements may come up in this context, as well as the rules for reaching agreement among shareholders; approval requirements and reserved matters; guidelines for management of the venture; and a joint plan for exit scenarios ranging from sale to third parties to an IPO.

"Sponsors embarking on a joint venture to develop a hydrogen project will be keen to understand the scope of their involvement. This may be influenced by the wider aims of the project, such as to break into a new market, gain knowledge or beat the competition."

5. Project financing

Project finance remains the most common way of financing renewables projects. In essence, project finance is a form of private debt financing of individual, usually large-scale, projects based on non-recourse or limited recourse funding. Rather than being tied to the cashflow of the project, payment of the sponsors is based upon the ultimate projected revenue of the completed power project. For this reason, project finance is considered to be off-balance sheet – one of its main attractions for sponsors in that sponsors and developers do not necessarily need sizeable pockets in order to secure funding.

Nevertheless, certainty for investors (whether sponsors or developers) in a project can be a challenge. This is due to the fact that project finance is heavily influenced by a combination of risk – of the specific project itself but also risks associated with the jurisdiction – and the project proposal, instead of a completed project. In order to achieve a level of certainty acceptable to the investors, extensive due diligence, financial modelling and revenue and cost forecasting have to be undertaken, amongst other things. Due to these lengthy procedures, project financing is a complex, time-consuming and administratively burdensome method of funding a power project.

The terms of financing agreements are in development, but are expected to follow those seen in project and corporate finance transactions in the wider renewables market. Given that project finance is usually more expensive than equity funding and takes considerably more time to arrange and document, it is likely that hydrogen projects will be equity-financed for some years yet, while the market grows and commercial viability is tested. This may be different in jurisdictions with a clear government mandate (and government backing of key risks) where developers of hydrogen projects may be compelled to choose debt financing instead.

A typical contractual structure for a hydrogen-powered project financing transaction would look something like Figure 7, with the SPV (Hydrogen Project Co) holding the key rights and being responsible for the debt.

Under the terms of such arrangements, the SPV will be the named contractor under any concession (or project) agreement and the recipient of the funding, as well as the counterparty employing construction, operation and maintenance services, the purchaser of fuel and the generator of power.

Figure 7. Typical contractual structure for a hydrogen-powered project financing transaction

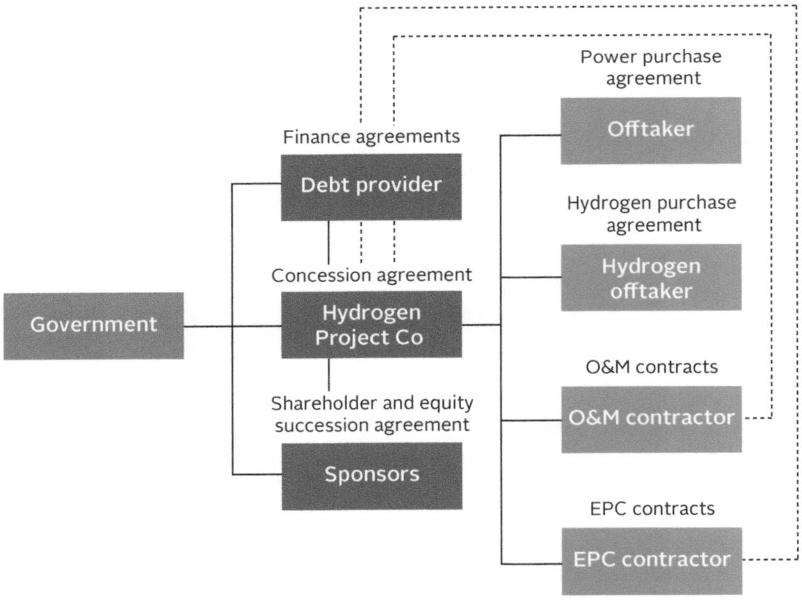

Source: Author.

As noted below, the key issue for hydrogen projects at present is cost. It is unlikely that major developments in the hydrogen sector (especially green or blue hydrogen) will succeed in the short term without government, or a quasi-government body, providing a subsidy or other regime from which the hydrogen project can benefit.

As the sector develops, and costs, including those of developing hydrogen projects, fall, so too will government involvement and support. For now, though, aside from the relevant government, the key parties with whom the SPV will contract will be its equity funders (ie, the sponsors) who have invested capital into the power project with the expectation of receiving a return once the project is operational. The key terms of the equity agreements and share subscription will set out how funding to the project is to be provided and the rights and benefits of each sponsor. In this context, the level of funding that the SPV will receive is likely to be heavily influence by the expertise of the sponsor. An experienced, well-connected sponsor is more likely to be able to guarantee hydrogen offtake, or a low-cost, stable supply of power, and so lower the risk of the lender not being paid back. Furthermore, lenders are more likely to give favourable terms under off-balance sheet mechanisms to sponsors they deem to be lower risk.

6. Leasing model of financing

Leasing is another form of financing model that may be suitable for hydrogen projects – in particular, it is emerging as a possible model for hydrogen-powered transport. Leasing offers an alternative where a project developer may not have the capital/equity to cover the upfront expenses or the means to develop the associated assets or infrastructure. While hydrogen technologies are still in early stages and costs are comparatively high against other established forms of renewable energy, leasing is an attractive business model for hydrogen projects.

In the last few years, several companies have been established with the aim of leasing hydrogen distribution and dispensing assets to transport businesses, for example Ryze Hydrogen. In its own words, Ryze Hydrogen aims to be a one-stop shop for affordable, green hydrogen equipment and infrastructure.[136] It owns – and will install and operate for its customers – hydrogen electrolysers, distribution networks and dispensing equipment. By removing the associated costs of developing and maintaining hydrogen-related assets, the leasing model becomes a viable option, particularly for government bodies. Ryze Hydrogen's model includes a tube trailer-based system that stores the compressed hydrogen, removing the need for permanent hydrogen storage, reducing costs and allowing the infrastructure to be brought to the customer. Ryze Hydrogen will use its equipment for hydrogen transportation as part of Transport for London's hydrogen double-decker bus fleet investment.[137] In another example, Toyota leases its hydrogen fuel cell passenger car, the Mirai, to Japanese government departments for business and industrial purposes.

In contrast to equity financing, where the project secures debt funding, the lenders will typically expect to have greater rights than the sponsor group and to be paid out ahead of equity funders. Typically in the renewables market, senior debt lenders tend to be financial institutions and development banks, such as the European Investment Bank or the World Bank. Subordinated debt tends to come from commercial lenders, these being a combination of institutional investors (pension funds, hedge funds and government bonds).

The attraction of lending to renewable projects has rocketed in recent years, with all manner of lenders now preferring to lend to green energy sources.

7. Opportunities and obstacles for investors and financiers

Although the opportunities of a hydrogen economy are often cited, there are nearly as many obstacles for sponsors and financiers to overcome, chief amongst them being:

- the lack of a stable regulatory framework;
- the requirement for subsidy to overcome current cost hurdles; and
- the need to ensure coordination of supply and demand.

Case study: EU gap-filling payments

The European Commission has, as of 1 January 2021, introduced an Energy Financing Mechanism as part of its Clean Energy Package. In its simplest form, the mechanism allows EU member states to choose to make financial contributions ('gap-filling' payments) towards renewable energy projects in other member states, where the contributing member state may be making slow progress to meeting their own renewable energy generation targets. The payments made will then contribute towards both member states' targets (though not distributed equally between them). Member states may also make additional payments to further the objective of the mechanism, separate from their progress to achieving their generation targets. As a whole, the mechanism aims to assist EU member states to meet their renewable energy generation targets while simultaneously mitigating against the cost of capital acting as a barrier to investment in renewable energy, particularly new technologies.

Payments made under the mechanism will be allocated through competitive tender with funding from payments generally being allocated to the lowest-cost projects. This may mean that low-carbon hydrogen projects are not initially favoured for funding, losing out to more developed and cost-efficient renewable technologies. However, at each funding round, the Commission may choose to allocate funding to particular projects or technologies, which may be beneficial for low-carbon hydrogen and associated technologies. Further, funding for projects under the mechanism may take the form of low-interest loans or grants, which would be particularly attractive for small-scale or pilot hydrogen projects.

Case study: state funding of low-carbon hydrogen projects in Japan

Japan is a key market to consider when it comes to state funding of low-carbon hydrogen projects. Having set a Basic Hydrogen Strategy back in 2017 that aims to cut CO_2 emissions by 80% by 2050 and decrease the price of hydrogen to have the same cost competitiveness as traditional energy sources,[138] the Japanese government has since established several routes for hydrogen projects to receive funding. The New Energy and Industrial Technology Development Organisation (NEDO) and Japan Oil, Gas and Metals National Corporation (JOGMEC) are two government organisations that offer funding for research, feasibility studies and pilot hydrogen projects. In recent years, NEDO has provided funding for several major hydrogen pilot projects including the Fukushima Hydrogen Energy Research Field (FH2R) which opened in Namie in March 2020. FH2R is the largest renewable hydrogen production unit in the world (10MW), using a power-to-grid system to power electrolysers that can produce 1,200 Nm3 of hydrogen per hour. NEDO is currently contributing to the ¥19 billion national project for the Development of Technologies for Realising a Hydrogen Society, as well as supporting hydrogen projects overseas, including the Latrobe Valley Project in Victoria, Australia. Financing is also widely available for commercial-scale hydrogen projects where Japanese companies play significant roles (as sponsors, contractors or offtakers) from institutions such as Japan Bank for International Cooperation, Nippon Export and Investment Insurance and also JOGMEC.

Subsidies and incentives have also been made available by the Japanese government for consumers wishing to purchase FCVs.

8. The cost barrier

The significant costs of developing both hydrogen projects and the infrastructure necessary for them to work at commercial scale are the key challenges faced by investors and financiers. The costs of low-carbon compared to grey hydrogen means that there are almost no applications at present where low-carbon hydrogen is a cost-effective fuel or feedstock. This means that without incentives for investment in the form of government support, the likelihood of low-carbon hydrogen developing at pace and scale, and thus achieving the necessary cost reductions, is doubtful. However, the clear driver for the uptake of low-carbon hydrogen, at least in jurisdictions with clear net zero commitments, is the decarbonisation opportunities it can offer.

Presently, global prices for hydrogen production range between $3.20–$7.70/kg where produced by renewables, $1.20–$2.10/kg where produced by natural gas with CCS and $0.70–$1.60/kg for grey hydrogen, according to figures produced in 2020 by the IEA.[139] However, while it may appear that blue hydrogen production is not that much more costly than grey, the commercialisation in cost terms of CCUS is a factor that needs to be taken into account. That said, there are reports that renewable hydrogen will be competitive with industrial supplies made from fossil fuels within the next 10 years in both Germany and the United States – contingent on political decisions about subsidies and tax credits.[140] This makes investments in blue and green hydrogen, at least in these jurisdictions, more attractive.

Case study: tax for a hydrogen economy in the United States[141]
Tax has played, and is expected to continue to play, an important role in the development of the hydrogen economy in the United States. At present, a number of federal tax incentives exist that specifically apply to hydrogen-related projects. These include:

- an investment tax credit (ITC) equal to 30% of the cost of investment in fuel cell power plants generating electricity;
- a credit for individual taxpayers for the cost of residential fuel cells;
- a credit for purchasers of alternative-fuel vehicles; and
- an ITC equal to 30% of the cost of alternative fuel vehicle refuelling property.

And – very relevant to blue hydrogen – there is a tax credit per tonne of carbon dioxide captured and sequestered.

With the publication of Joe Biden's American Jobs Plan at the end of March 2021,[142] it looks likely that hydrogen will (and will be lobbied to) play a role in forthcoming clean energy tax incentives. The plan, which proposes to invest over $2 trillion into infrastructure, emphasises Biden's ambitions to re-energise the United States' power infrastructure on the path to achieving 100% carbon-free electricity by 2035. In doing so, it details the president's aims to mobilise private investment for a modern power sector.

Specifically, it proposes a ten-year extension and phase-down of an expanded ITC and production tax credit (PTC) for clean energy generation and storage. A new PTC for hydrogen projects in distressed communities is also proposed, alongside funding for fifteen decarbonised hydrogen demonstration projects, although details of the form or enactment of these proposals are not yet specified.

Notwithstanding the cost challenges, the pace at which renewable electricity generation projects are developing and becoming cheaper also offers the prospect of low-carbon hydrogen projects following suit. At present, energy generation (from renewable sources) accounts for around 80% of the cost of hydrogen produced by electrolysis.[143] As the cost of renewable generation falls, and as electrolysers get bigger and cheaper, the hurdle of the cost of producing low-carbon hydrogen should, subject to support from governments to kick-start the projects, also diminish.

9. Getting supply to meet demand

There is little point in investing in hydrogen production when the demand cannot be guaranteed. Equally, how can an investor finance projects reliant on hydrogen when the supply is not sufficiently mature

"As the cost of renewable generation falls, and as electrolysers get bigger and cheaper, the hurdle of the cost of producing low-carbon hydrogen should, subject to support from governments to kick-start the projects, also diminish."

or reliable? This chicken and egg dilemma is common to new start-ups and technologies, and low-carbon hydrogen is no exception.

As a result of this uncertainty, investors try to create artificial barriers and risk transfers which push up the prices of their investments, and government intervention is key to creating that initial pathway in nearly all jurisdictions that will allow further investment to flow. In Europe, the European Union announced the €463 billion Hydrogen Strategy in July 2020 as part of its wider Green Deal.[144]

Since the 2015 Paris Agreement, an increasing driver for investment into renewable, including low-carbon, technologies, has been the rise of responsible investment – ESG (environmental, social, governance) investing.[145]

Table 3: Components of ESG investing

Component	Categories
Environmental	Climate risks and mitigation
	Resource change, scarcity and management
	Low-carbon energy
Social	Diversity
	Human rights
	Security
Governance	Openness and transparency
	Ethics and values
	Eliminating corruption

While the concept of ESG investing was first described by then-UN Secretary-General Kofi Annan in 2004,[146] it has taken a couple of decades to get onto the decision-making radar of asset managers and institutional investors, stimulated in part by the rise in the socially responsible investment movement more broadly.

Increasingly, ESG considerations are being integrated into the charters of global companies and thus provide an impetus for them to invest in renewable and sustainable technologies, of which hydrogen is one.

"Since the 2015 Paris Agreement, an increasing driver for investment into renewable, including low-carbon, technologies, has been the rise of responsible investment."

The value of global assets applying ESG data to drive investment decisions has more than tripled over the last eight years to $40.5 trillion in 2020.[147] Already in 2017, KPMG was reporting that ESG investments accounted for about one-quarter of all professionally managed investments globally.[148]

For hydrogen projects, a key opportunity in seeking investment is the clarity recently introduced around what counts as ESG investment. For example, the European Union has developed a common classification system to identify whether certain economic activities are 'environmentally sustainable'. As part of the European Commission's Sustainable Finance Action Plan, financial market participants (such as asset managers and institutional investors) and financial advisers are required by the Sustainable Finance Disclosure Regulation (SFDR) to make specific disclosures in relation to their ESG considerations.[149] The aim of the SFDR is to provide transparency for investors with regard to what their investment is ultimately funding, highlighting whether it is a 'sustainable investment' and looking at 'sustainability risk' and 'sustainability factors'. As set out in Table 4, financial institutions are required to make certain disclosures on their websites, provide pre-contractual disclosures in certain instances, and produce periodic reports on their ESG commitments.[150]

Table 4. Sustainability-related disclosures and reporting obligations under SFDR[151]

Requirement	Financial market participant	Financial adviser	Type of financial product
Website disclosures			
Sustainability risk policies (Article 3)	Y	Y	All (within scope of the SFDR)
Adverse sustainability impacts (Article 4)	Y	Y	All
Remuneration policies (Article 5)	Y	Y	All
Promoting environmental or social characteristics or sustainable investments (Article 10)	Y	N	Financial product that promotes environmental or social characteristics Financial product that pursues sustainability objectives
Pre-contractual disclosures			
Sustainability risks (Article 6)	Y	Y	All
Adverse sustainability impacts (Article 7)	Y	N	All
Promoting environmental or social characteristics (Article 8)	Y	N	Financial product that promotes environmental or social characteristics
Sustainable investments (Article 9)	Y	N	Financial product that pursues sustainability objectives
Periodic reports			
Promoting environmental or social characteristics or sustainable investments (Article 11)	Y	N	Financial product that promotes environmental or social characteristics Financial product that pursues sustainability objectives

Source: Table based on information in Practical Law, "Sustainable finance: SFDR: overview" (note 151).

With the rise of sustainable finance, hydrogen projects are becoming increasingly attractive to investors and lenders – whether European institutions seeking to diversify into this emerging market or private investors who choose to fund certain categories of emerging technologies as a matter of corporate responsibility. As an example, in providing funding (in whatever form) to a producer of green hydrogen, an EU financial institution is able to disclose that it has considered Article 4 (adverse sustainability impacts), and Articles 8 and 10 (promoting environmental or social characteristics or sustainable investments) of the SFDR, though to what extent will depend on the level of funding provided.

ESG considerations are beginning to shape thinking in non-European jurisdictions too. Though not at a national level, in Singapore the Overseas-Chinese Banking Corporation (OCBC Bank) has set itself a target of 2025.[152] Renewable energy and clean transportation, among other sustainable finance sectors, are considered by the OCBC Bank to have high growth potential across the Asia-Pacific region and have seen a significant increase in sustainable financing in recent years.

While it was once thought the United Kingdom would directly implement the SFDR, the UK government published its own roadmap for implementing mandatory climate disclosures aligned with the Task Force on Climate-related Financial Disclosures (TCFD). Many of the rules are set to come into force by 2023, with disclosure to become fully mandatory across the economy by 2025.[153]

The EU Hydrogen Strategy[154]

The EU Hydrogen Strategy is predicated on driving down the cost of producing green hydrogen in Europe, currently between 2.5 and 5.5 euros per kg, to between 1.1 and 2.4 euros per kg by 2030. Given that electrolysers are one of the four main cost components (the others being renewable electricity, the capacity factor at which plants run and cost of capital), this is a significant incentive for European hydrogen projects.

As the cost of renewable electricity generation continues to fall, so has the cost of electrolysers (though not as fast!). There are reports that Chinese manufacturers are already capable of achieving the EU's aims, supplying equipment at $200/kW. Meanwhile the Hydrogen Strategy aims to drive down the cost of electrolysers from €900/kW to €450/kW or less in the period after 2030. Besides the costs, however, a question remains whether the more responsive PEM technologies which are able to ramp up and down within tenths of a second would be favoured over the alkaline electrolysers which have been the focus of manufacturers in China.

Developers of hydrogen would also need to ensure that their electricity inputs can be secured at prices that make the production and sale of their hydrogen economic. This may be by choosing to buy electricity at times when it is effectively free or being co-located with renewable generation to take advantage of outputs from the renewable generation that cannot be exported to the grid. This is possible to imagine in a number of jurisdictions that are developing significant renewable capacity, such as parts of southern Europe, Australia, Morocco, the Middle East, Mexico, Chile, Brazil, southern United States, China and India.

VII. Conclusion

The unique characteristics of low-carbon hydrogen projects mean that legal and policy frameworks will need careful consideration to ensure that ambitions can be realised. Treating hydrogen like another gas or a renewable electricity source ignores the specific issues that working with this light, colourless and odourless gas present. Common barriers to current efforts to promote hydrogen development include complex laws such as the European agreement concerning ADR,[155] which place very high safety requirements on hydrogen transport and storage, even when undertaken in relatively modest volumes.

In order to create an appropriate investment case, the financing structures, operational requirements and revenue stream arrangements, as well as a range of other elements, need to be considered to present an effective commercialisation model. Although the chemical element may fundamentally be the same, how and where it is used will vary and it will not always slot neatly into what wider legal frameworks encourage. For example, the 2020 EU taxonomy limits 'low-carbon' or 'renewable' hydrogen to blue and green by virtue of requirements that:[156]

- the level of direct CO_2 emissions from the manufacture of hydrogen is less than 5.8 tCO2 e/t;

"Whether hydrogen can play a significant part in the energy transition will depend on the alignment of clear legislative and regulatory frameworks with supportive government schemes that incentivise and protect innovative investments, on the one hand, and the cooperation of a joined-up and collaborative industry on the other."

- electricity use for hydrogen produced by electrolysis is at or lower than 58 MWh/t; and
- average carbon intensity of the electricity produced for hydrogen manufacturing is at or below 100 gCO_2 e/kWh.

Whether hydrogen can play a significant part in the energy transition will depend on the alignment of clear legislative and regulatory frameworks with supportive government schemes that incentivise and protect innovative investments, on the one hand, and the cooperation of a joined-up and collaborative industry on the other. And whether individual jurisdictions can reap the undoubted rewards of bold legislative steps to create new, integrated frameworks which holistically address the role that hydrogen is expected to play in tomorrow's energy systems will depend on the relevant political regimes and their appetite for change.

Experience with other emerging technologies tells us that living with uncertainty and lack of guidance results in inconsistencies of approach and protracted (and costly) negotiations as investors, developers and authorities navigate the not-quite-fit-for-purpose rules. And cost reduction is the goal here – to achieve the IEA's prediction that the cost of producing hydrogen from renewable electricity should fall by 30% by 2030 on the back of declining costs of renewables and the scaling-up of production. Coordination between governments, industry and investors to address the existing regulatory and legal barriers and create a common framework that facilitates the development of harmonised standards for the production, transportation and storage of hydrogen, and for tracing the environmental impacts of the hydrogen supplied, are key.[157]

Without innovation, combined with steady demand to meet the developing supply, the opportunities presented by hydrogen as an energy vector may never be seized.

Notes

1 The diagram is based on Fig 1 in Christopher J Quarton and Sheila Samsatli, "Power-to-gas for injection into the gas grid: What can we learn from real-life projects, economic assessments and systems modelling?", ScienceDirect, *Renewable and Sustainable Energy Reviews*, vol 98, December 2018, pp302–316. Available at: www.sciencedirect.com/science/article/pii/S1364032118306531.

2 "Hydrogen", IEA report, May 2021. Available at: www.iea.org/fuels-and-technologies/hydrogen.

3 Committee on Climate Change, "Hydrogen in a low-carbon economy", November 2018. Available at: www.theccc.org.uk/wp-content/uploads/2018/11/Hydrogen-in-a-low-carbon-economy.pdf.

4 Department for Business, Energy & Industrial Strategy (BEIS), "Powering our Net Zero Future", white paper, December 2020. Available at: https://assets.publishing.service.gov.uk/government/uploads/system/uploads/attachment_data/file/945899/201216_BEIS_EWP_Command_Paper_Accessible.pdf.

5 UK Research and Innovation, "UKRI awards £171m in UK decarbonisation to nine projects", news announcement, 17 March 2021. Available at: www.ukri.org/news/ukri-awards-171m-in-uk-decarbonisation-to-nine-projects/.

6 See: BEIS, "Carbon Capture, Usage and Storage – A Government Response", August 2020, available at https://assets.publishing.service.gov.uk/government/uploads/system/uploads/attachment_data/file/909706/CCUS-government-response-business-models.pdf; BEIS, "Carbon capture, usage and storage (CCUS): business models" (in particular section 6), 21 December 2020, available at www.gov.uk/government/publications/carbon-capture-usage-and-storage-ccus-business-models; and BEIS, "Business models for low carbon hydrogen production", 17 August 2020, available at www.gov.uk/government/publications/business-models-for-low-carbon-hydrogen-production.

7 European Commission, "Powering a climate-neutral economy: Commission sets out plans for the energy system of the future and clean hydrogen", press announcement, 8 July 2020. Available at: https://ec.europa.eu/commission/presscorner/detail/en/ip_20_1259.

8 European Commission, "A hydrogen strategy for a climate-neutral Europe", communication to the European Parliament, Council, Economic and Social Committee and Committee of the Regions, 8 July 2020. Available at: https://ec.europa.eu/energy/sites/ener/files/hydrogen_strategy.pdf.

9 European Commission, "Powering a climate-neutral economy", *supra*.

10 "Shell opens 10 MW German hydrogen electrolyser to boost green fuel output", *Reuters*, 2 July 2021. Available at: www.reuters.com/business/energy/shell-opens-10-mw-german-hydrogen-electrolyser-boost-green-fuel-output-2021-07-02/.

11 Hydrogen Council and McKinsey & Company, "Hydrogen Insights – A perspective on hydrogen investment, market development and cost competitiveness", February 2021. Available at: https://hydrogencouncil.com/wp-content/uploads/2021/02/Hydrogen-Insights-2021.pdf.

12 The graph is from "Hydrogen: A Renewable Energy Perspective", International Renewable Energy Agency report, 2019. Available at: www.irena.org/-/media/Files/IRENA/Agency/Publication/2019/Sep/IRENA_Hydrogen_2019.pdf.

13 Department for Business, Energy & Industrial Strategy, "The Clean Growth Strategy: Leading the way to a low carbon future", October 2017. Available at: https://assets.publishing.service.gov.uk/government/uploads/system/uploads/attachment_data/file/700496/clean-growth-strategy-correction-april-2018.pdf.

14 Committee on Climate Change, "Hydrogen in a low-carbon economy", *supra*.

15 "Hydrogen: A Renewable Energy Perspective", report by the International Renewable Energy Agency, *supra*.

16 For the purpose of this Special Report, we will use the broad term 'hydrogen' to refer to low-carbon hydrogen (ie, green and blue hydrogen) unless otherwise specifically indicated.

17 "Analysis: do hydrogen-powered cars have a future?", *Autocar*, 13 November 2019. Available at: www.autocar.co.uk/car-news/industry/analysis-do-hydrogen-powered-cars-have-future.

18 Azusa Kawakami, "Mitsubishi Heavy to build biggest zero-carbon steel plant", *Nikkei Asia*, 28 December 2020, available at: https://asia.nikkei.com/Spotlight/Environment/Climate-Change/Mitsubishi-Heavy-to-build-biggest-zero-carbon-steel-plant; "Hydrogen", IEA report, *supra*; and EU-Japan Centre, "Hydrogen Steel Plant: Voestalpine X Mitsubishi Heavy Industries", case study, 2021, available at: www.eu-japan.eu/publications/hydrogen-steel-plant-voestalpine-x-mitsubishi-heavy-industries.

19 European Commission, "Heating and cooling", 11 March 2021. Available at: https://ec.europa.eu/energy/topics/energy-efficiency/heating-and-cooling_en?redir=1.

20 "Hydrogen: A Renewable Energy Perspective", IRENA, *supra*.

21 Hydrogen Council, "Hydrogen – scaling up: A sustainable pathway for the global energy transition", November 2017. Available at: https://hydrogencouncil.com/wp-content/uploads/2017/11/Hydrogen-Scaling-up_Hydrogen-Council_2017.compressed.pdf.

22 See: UNFCCC, "What is the United Nations Framework Convention on Climate Change?". Available at: https://unfccc.int/process-and-meetings/the-convention/what-is-the-united-nations-framework-convention-on-climate-change.

23 See: UNFCCC, "What is the Paris Agreement?". Available at: https://unfccc.int/process-and-meetings/the-paris-agreement/the-paris-agreement.

24 See: UNFCCC, "The Paris Agreement and NDCs". Available at: https://unfccc.int/process-and-meetings/the-paris-agreement/nationally-determined-contributions-ndcs/nationally-determined-contributions-ndcs.

25 Government of Chile, "Chile's nationally determined contribution – Update 2020". Available at: www4.unfccc.int/sites/ndcstaging/PublishedDocuments/Chile%20First/Chile%27s_NDC_2020_english.pdf.

26 See: UNFCCC, "What is the CDM?". Available at: https://cdm.unfccc.int/about/index.html.

27 See: UNFCCC, "Kyoto Protocol – Targets for the first commitment period". Available at: https://unfccc.int/process-and-meetings/the-kyoto-protocol/what-is-the-kyoto-protocol/kyoto-protocol-targets-for-the-first-commitment-period.

28 For the full text of the Paris agreement, see: https://unfccc.int/sites/default/files/english_paris_agreement.pdf.

29 For more on the bottom-up approach of Article 6.2, see Axel Michaelowa, Sonja Butzengeiger, Romain Debarre, Adnan Shihab-Eldin, Richard Forrest, Claude Mandil and Antoine Rostand, "Promoting carbon-neutral hydrogen through UNFCCC and national-level policies", G20 Insights, 22 November 2020. Available at: www.g20-insights.org/policy_briefs/promoting-carbon-neutral-hydrogen-through-unfccc-and-national-level-policies/.

30 For more on the Green Hydrogen Catapult, see: Climate Champions, "Green Hydrogen Catapult: World's green hydrogen leaders unite to drive 50-fold scale-up in six years", Race to Zero campaign, 8 December 2020. Available at: https://racetozero.unfccc.int/green-hydrogen-catapult/.

31 See: European Commission, "A hydrogen strategy for a climate-neutral Europe", supra.

32 See: CMS, Hydrogen Law and Regulation in the Netherlands, CMS Expert Guides. Available at: https://cms.law/en/int/expert-guides/cms-expert-guide-to-hydrogen/netherlands.

33 The Climate Agreement is available from the Dutch Government's website at: www.government.nl/binaries/government/documents/reports/2019/06/28/climate-agreement/Climate+Agreement.pdf.

34 See: Overheid.nl (portal for information on government organisations in the Netherlands), Regeling gaskwaliteit. Available at: https://wetten.overheid.nl/BWBR0035367/2016-04-01 (Dutch language only).

35 See: Clifford Chance, "Focus on Hydrogen: Portugal's Clean Energy Plan Gathers Pace", briefing, 16 November 2020, available at: www.cliffordchance.com/content/dam/cliffordchance/briefings/2020/11/focus-on-hydrogen-portugals-clean-energy-plan-gathers-pace.pdf; and Vieira de Almeida, "A Simplified Guide to the Portuguese Hydrogen Strategy", available at: www.vda.pt/xms/files/05_Publicacoes/2020/Flashes_Newsletters/VdA_-_Guide_to_the_Portuguese_Hydrogen_Strategy.pdf.

36 Available at: https://ec.europa.eu/energy/sites/ener/files/documents/pt_final_necp_main_en.pdf.

37 Available at: https://kig.pl/wp-content/uploads/2020/07/EN_H2_ENG.pdf.

38 Government of the Netherlands, "Memorandum of Understanding between the Minister of Environment and Climate Action of the Portuguese Republic and the Minister of Economic Affairs and Climate Policy of the Netherlands in the field of Energy – Hydrogen", 17 August 2020. Available at: www.government.nl/binaries/government/documents/publications/2020/09/23/memorandum-of-understanding-between-the-netherlands-and-portugal-concerning-green-hydrogen/MoU_PT-NL+H2_signed.pdf.

39 See: Lexology, "A Simplified Guide to the Portuguese Hydrogen Strategy", 26 November 2020. Available at: www.lexology.com/library/detail.aspx?g=1492dd16-64f6-4d7a-a0eb-c88bce3b975f.

40 Great Britain comprises England, Wales and Scotland and is regulated separately from Northern Ireland. The legal systems are somewhat different as between England and Wales, Scotland, and Northern Ireland.

41 Paul E Dodds, Anthony Velazquez Abad, Will McDowall and Gerard I Fox, "Opportunities for hydrogen and fuel cell technologies to contribute to clean growth in the UK", H2FC Supergen white paper, May 2020. Available at: www.h2fcsupergen.com/wp-content/uploads/2020/04/2020_04_H2FC_Supergen_Hydrogen_Fuel_Cells_P_Dodds_DIGITAL_W_COVER_v05.pdf.

42 IEA, "The Future of Hydrogen: Seizing today's opportunities", June 2019. Available at: www.iea.org/reports/the-future-of-hydrogen.

43 Capgemini Invent with Breakthrough Energy, "Fit for Net-Zero – 55 Tech Quests to accelerate Europe's recovery and pave the way to climate neutrality", report, October 2020. Available at: www.capgemini.com/wp-content/uploads/2020/10/Net-zero-main-report-2020.pdf.

44 See: Ofgem, Renewables Obligation (RO). Available at: www.ofgem.gov.uk/environmental-and-social-schemes/renewables-obligation-ro.

45 Spenser J Robinson, Robert Allan Simons, Eunkyu Lee and Andrew C Kern, "Demand for Green Buildings: Office Tenants' Stated Willingness-to-Pay for Green Features", Journal of Real Estate Research, vol 38 no 3, July 2016. Available from: www.researchgate.net/publication/286124374_Demand_for_Green_Buildings_Office_Tenants%27_Stated_Willingness-to-Pay_for_Green_Features.

46 Rich McEachran, "Growing need for greener buildings", Raconteur, 1 March 2021. Available at: www.raconteur.net/infrastructure/growing-need-for-greener-buildings/.

47 See the Clean Energy Finance Corporation website: www.cefc.com.au/where-we-invest/special-investment-programs/advancing-hydrogen-fund/.

48 Australian Renewable Energy Agency, "Seven shortlisted for $70 million hydrogen funding round". Available at: https://arena.gov.au/news/seven-shortlisted-for-70-million-hydrogen-funding-round/.

49 Hydrogen Europe, "Green Hydrogen Investment and Support Report: Hydrogen Europe's input for a post-COVID-19 recovery plan". Available at: http://profadvanwijk.com/wp-content/uploads/2020/05/Hydrogen-Europe_Green-Hydrogen-Recovery-Report_final.pdf.

50 For more on the JCM, see: www.mofa.go.jp/ic/ch/page1we_000105.html.

51 See: EU taxonomy for sustainable activities. Available at: https://ec.europa.eu/info/business-economy-euro/banking-and-finance/sustainable-finance/eu-taxonomy-sustainable-activities_en.

52 ITM Power, "€5m EU Award to Study Offshore Green Hydrogen Production with Ørsted and Siemens Gamesa". Available at: www.itm-power.com/news/5m-eu-award-to-study-offshore-green-hydrogen-production-with-orsted-and-siemens-gamesa.

53 See: www.fch.europa.eu/page/who-we-are.

54 The costs figures are taken from Pöyry Management Consulting, "Hydrogen from natural gas – the key to deep decarbonisation", July 2019. Available at: www.poyry.com/sites/default/files/media/related_material/zukunft_erdgas_key_to_deep_decarbonisation.pdf.

55 Martin Lambert, "Hydrogen and decarbonisation of gas: false dawn or silver bullet?", Insight paper for the Oxford Institute for Energy Studies, March 2020. Available at: www.oxfordenergy.org/wpcms/wp-content/uploads/2020/03/Insight-66-Hydrogen-and-Decarbonisation-of-Gas.pdf.

56 The table is adapted from Pöyry Management Consulting, "Hydrogen from natural gas – the key to deep decarbonisation", *supra*. Martin Lambert, in his Insight paper for the Oxford Institute for Energy Studies, "Hydrogen and decarbonisation of gas" (*ibid*), also notes that $2/kg equates to €45/MWh and $3/kg to €67/MWh, so the two analyses are broadly consistent.

57 See: European Commission, "Clean Energy for All Europeans – unlocking Europe's growth potential", November 2016, available at: http://europa.eu/rapid/press-release_IP-16-4009_en.htm; and "EU hydrogen strategy", available at: https://ec.europa.eu/energy/topics/energy-system-integration/hydrogen_en#eu-hydrogen-strategy.

58 Council Directive 2009/73/ec of 13 July 2009 concerning common rules for the internal market in natural gas and repealing Directive 2003/55/EC. Available at: https://eur-lex.europa.eu/legal-content/EN/TXT/?uri=CELEX%3A02009L0073-20190523.

59 Directive (EU) 2018/2001 of the European Parliament and of the Council of 11 December 2018 on the promotion of the use of energy from renewable sources (recast). Available at: https://eur-lex.europa.eu/legal-content/EN/TXT/?uri=uriserv:OJ.L_.2018.328.01.0082.01.ENG.

60 Ofgem, "Our powers and duties". Available at: www.ofgem.gov.uk/publications-and-updates/our-powers-and-duties.

61 See: Draft Amendment to Energy Act regarding the regulation of hydrogen networks, Lexology, 3 March 2021. Available at: www.lexology.com/library/detail.aspx?g=09970757-9647-490a-aebe-88aafc31aaab.

62 See: Gas for Climate, "European Hydrogen Backbone". Available at: https://gasforclimate2050.eu/sdm_downloads/european-hydrogen-backbone/.

63 See: "Extending the European Hydrogen Backbone: A European Hydrogen Infrastructure Vision Covering 21 Countries", Creos and others with Guidehouse, April 2021. Available at: https://gasforclimate2050.eu/sdm_downloads/extending-the-european-hydrogen-backbone/.

64 Gas for Climate, "European Hydrogen Backbone", *supra*.

65 See: "The Vision of HySTRA". Available at: www.hystra.or.jp/en/about/.

66 "The Future of Hydrogen: Seizing today's opportunities", IEA technology report, June 2019. Available at: www.iea.org/reports/the-future-of-hydrogen.

67 *Ibid.*

68 "Vattenfall aims for carbon-free gas power … Cooperation with Statoil and Gasunie started", Vattenfall newsroom, 23 April 2018. (Statoil became Equinor the following month.) Available at: https://group.vattenfall.com/press-and-media/newsroom/2017/vattenfall-aims-for-carbon-free-gas-power.

69 "Industry decarbonisation: HYBRIT, a collaboration between SSAB, LKAB and Vattenfall", Vattenfall website. Available at: https://group.vattenfall.com/what-we-do/roadmap-to-fossil-freedom/industry-decarbonisation/hybrit.

70 See: HyNet North West, "HyNet North West – Unlocking net zero for the UK". Available at: https://hynet.co.uk/wp-content/uploads/2020/10/HyNet_NW-Vision-Document-2020_FINAL.pdf.

71 See: "We don't stand for hot air around here", Net Zero Teesside website. Available at: www.netzeroteesside.co.uk/project/.

72 See: "Hydrogen Delta", Smart Delta Resources website. Available at: www.smartdeltaresources.com/en/hydrogen-delta.

73 See: "CO2 reduction through storage beneath the North Sea", Porthos website. Available at: www.porthosco2.nl/en/.

74 See: James Burgess, "Interview: UK's National Grid looks to prove case for hydrogen blending", S&P Global Platts, 29 April 2021. Available at: www.spglobal.com/platts/en/market-insights/latest-news/electric-power/042921-interview-uks-national-grid-looks-to-prove-case-for-hydrogen-blending.

75 "Tomorrow starts today: The Carbon2Chem® project", Thyssenkrupp group website. Available at: www.thyssenkrupp.com/en/newsroom/content-page-162.html.

76 See: "Ústí Region bets on hydrogen", Orlen Unipetrol press release, 19 November 2019. Available at: www.unipetrol.cz/en/Media/PressReleases/Pages/20191119_TZ_HYDROGENE_USTI_EN.aspx.

77 Department for Business, Energy and Industrial Strategy, "Clean Growth – Transforming Heating", December 2018. Available at: https://assets.publishing.service.gov.uk/government/uploads/system/uploads/attachment_data/file/766109/decarbonising-heating.pdf.

78 CMS, *CMS Expert Guide to Hydrogen Law and Regulation*. Available at: https://cms.law/en/int/expert-guides/cms-expert-guide-to-hydrogen.

79 International Council on Clean Transportation, "Hydrogen for heating? Decarbonization options for households in the United Kingdom in 2050", white paper. Available at: https://theicct.org/sites/default/files/publications/Hydrogen-heating-UK-dec2020.pdf

80 Planning, BIM & Construction Today (pbctoday), "Green light for 'world-first' hydrogen heating network", 30 November 2020. Available at: www.pbctoday.co.uk/news/energy-news/hydrogen-heating-network/85953/#.

81 "The GRHYD demonstration project", Engie website. Available at: www.engie.com/en/businesses/gas/hydrogen/power-to-gas/the-grhyd-demonstration-project.

82 Cordis, "Green Hysland – Deployment of an H2 Ecosystem on the island of Mallorca", available at: https://cordis.europa.eu/project/id/101007201; FCH, "Green Hysland in Mallorca, the first green hydrogen project in a Mediterranean country due to get European funding", available at: www.fch.europa.eu/news/green-hysland-mallorca-first-green-hydrogen-project-mediterranean-country-due-get-european; and "Enagás and ACCIONA launch Power to Green Hydrogen Mallorca project", *Bioenergy International*, 31 December 2020, available at: https://bioenergyinternational.com/storage-logistics/enagas-and-acciona-launch-power-to-green-hydrogen-mallorca-project.

83 See: Paul Mah, "KBR to study hydrogen-based energy supply for Keppel Data Centres", Data Centre Dynamics (DCD), 2 February 2021. Available at: www.datacenterdynamics.com/en/news/kbr-study-hydrogen-based-energy-supply-keppel-data-centres/.

84 See: "NUS Engineering, Keppel and SLNG develop new energy-efficient cooling technology for data centres", NUS Engineering. Available at: www.eng.nus.edu.sg/news/nus-engineeing-keppel-and-slng-develop-new-energy-efficient-cooling-technology-for-data-centres/.

85 See: Sambit Mohanty and Fred Wang, "Hydrogen-based data centers are in Singapore's decarbonized, digital economy vision", S&P Global Platts, 30 June 2020. Available at: www.spglobal.com/platts/en/market-insights/latest-news/electric-power/063020-hydrogen-based-data-centers-are-in-singapores-decarbonized-digital-economy-vision.

86 See: Sebastian Moss, "Keppel Data Centres signs another MoU for LNG and hydrogen floating data center project", Data Centre Dynamics (DCD), 26 October 2020. Available at: www.datacenterdynamics.com/en/news/keppel-data-centres-signs-another-mou-lng-and-hydrogen-floating-data-center-project/.

87 See: Department for Business, Energy and Industrial Strategy (BEIS) guidance on heat networks. Available at: www.gov.uk/guidance/heat-networks-overview.

88 Department for Business, Energy and Industrial Strategy, "Clean Growth – Transforming Heating", *supra*.

89 Phil Lattimore, "Fuel for thought – prototype hydrogen gas boilers", *CIBSE Journal*, January 2020.

90 For more on HyDeploy see: https://hydeploy.co.uk/.

91 See: CMS, *Hydrogen Law and Regulation in the Netherlands*, *supra*, p116.

92 See: CMS, *Hydrogen Law and Regulation in Portugal*, p133. Available at: https://cms.law/en/int/expert-guides/cms-expert-guide-to-hydrogen/portugal.

93 George Varma and Tim Armsby, "Clean hydrogen: building a sustainable market in the Middle East", Pinsent Masons analysis, 20 January 2021. Available at: www.pinsentmasons.com/out-law/analysis/clean-hydrogen-building-sustainable-market-middle-east.

94 CMS, *CMS Expert Guide to Hydrogen Law and Regulation*, *supra*, p12.

95 Clifford Chance, "Focus on Hydrogen: A €7.2 billion strategy for hydrogen delivery in France", www.cliffordchance.com/content/dam/cliffordchance/briefings/2020/10/focus-on-hydrogen-eur-7-2-Billion-strategy-for-hydrogen-energy-in-france.pdf.

96 Norwegian Ministry of Petroleum and Energy and Norwegian Ministry of Climate and Environment, "The Norwegian Government's hydrogen strategy", strategy paper. Available at: www.regjeringen.no/contentassets/8ffd54808d7e42e8bce8134ob13b6b7d/hydrogenstrategien-engelsk.pdf.

97 "Comparison of hydrogen and battery electric trucks", *Transport & Environment*, June 2020. Available at: www.transportenvironment.org/sites/te/files/publications/2020_06_TE_comparison_hydrogen_battery_electric_trucks_methodology.pdf.

98 See: "Jo Bamford: 'Wrightbus ready to produce 3,000 hydrogen buses for UK by 2024'", *Sustainable Bus*, 28 April 2020. Available at: www.sustainable-bus.com/news/jo-bamford-wrightbus-3000-hydrogen-buses/.

99 See: "Hydrogen Economy Development in Korea", Netherlands Enterprise Agency, June 2020, available at: www.rvo.nl/sites/default/files/2020/07/Korea-Hydrogen-economy-overview-2020-final.pdf; and Clifford Chance, "Focus on hydrogen: Korea's new energy roadmap", October 2020, available at: www.cliffordchance.com/content/dam/cliffordchance/briefings/2020/10/focus-on-hydrogen-korea-new-energy-roadmap.pdf.

100 "Korean hydrogen vehicles supply highest in the world", *Big News Network*. Available at: www.bignewsnetwork.com/news/269088631/korean-hydrogen-vehicles-supply-highest-in-the-world.

101 See: Ministry of Economic Affairs and Climate Policy, the Netherlands, "Hydrogen Economy Development in Korea", 12 June 2020. Available at: www.rvo.nl/sites/default/files/2020/07/Korea-Hydrogen-economy-overview-2020-final.pdf.

102 Woodside, "Hydrogen demand", 11 January 2021. Available at: www.woodside.com.au/media-centre/news-stories/story/hydrogen-demand.

103 See: European Agreement Concerning the International Carriage of Dangerous Goods by Road (ADR), 1 January 2017. Available at: https://unece.org/DAM/trans/danger/publi/adr/adr2017/ADR2017E_web.pdf.

104 See: "Deployment of Hydrogen Fuel Cell Trucks", *Energy Industry Review*, 6 March 2020. Available at: https://energyindustryreview.com/events/deployment-of-hydrogen-fuel-cell-trucks/.

105 See: "Coalition Statement on the deployment of fuel cell and hydrogen heavy-duty trucks in Europe", November 2020. Available at: www.fch.europa.eu/sites/default/files/FCH%20Docs/201215_Coalition%20Statement%20on%20deployment%20of%20FCH%20trucks%20in%20Europe.pdf.

106 "The JIVE projects celebrate a milestone: over 200 fuel cell buses have been ordered!", Fuel Cell Electric Buses knowledge base, 5 October 2020. Available at: www.fuelcellbuses.eu/public-transport-hydrogen/jive-projects-celebrate-milestone-over-200-fuel-cell-buses-have-been.

107 For more on H2ME 1 and H2ME 2 see: https://h2me.eu/about/.

108 See: "Coradia iLint – the world's 1st hydrogen powered train". Available on the Alstom website at: www.alstom.com/solutions/rolling-stock/coradia-ilint-worlds-1st-hydrogen-powered-train.

109 "UK's First Hydrogen Train takes to the Mainline", Porterbrook, 30 September 2020. Available at: www.porterbrook.co.uk/news/uks-first-hydrogen-train-takes-to-the-mainline.

110 See: IMO, "Initial IMO GHG Strategy". Available at: www.imo.org/en/MediaCentre/HotTopics/Pages/Reducing-greenhouse-gas-emissions-from-ships.aspx.

111 Randall Krantz, Kasper Søgaard and Dr Tristan Smith, "The scale of investment needed to decarbonize international shipping", Global Maritime Forum insight brief, January 2020. Available at: www.globalmaritimeforum.org/news/the-scale-of-investment-needed-to-decarbonize-international-shipping.

112 Shell, "Decarbonising Shipping: Setting Shell's Course". Available at: www.shell.com/promos/energy-and-innovation/decarbonising-shipping-setting-shells-course/_jcr_content.stream/1601385103966/709d83f692075a4f1880104fc5cc466168e8a26a/decarbonising-shipping-setting-shells-course.pdf.

113 Ministry of Petroleum and Energy and Ministry of Climate and Environment, "The Norwegian hydrogen strategy", press release, 8 June 2020. Available at: www.regjeringen.no/en/aktuelt/the-norwegian-hydrogen-strategy/id2704774/.

114 "Hydrogen-powered aviation: A fact-based study of hydrogen technology, economics, and climate impact by 2050", report by McKinsey & Company for Clean Sky 2 JU and FCH 2 JU, May 2020. Available at: www.fch.europa.eu/sites/default/files/FCH%20Docs/20200507_Hydrogen%20Powered%20Aviation%20report_FINAL%20web%20%28ID%208706035%29.pdf.

115 See: Clean Sky, "EU to set up a new European Partnership for Clean Aviation", 23 February 2021. Available at: www.cleansky.eu/news/eu-to-set-up-a-new-european-partnership-for-clean-aviation.

116 See: "Jo Bamford: 'Wrightbus ready to produce 3,000 hydrogen buses", *supra*.

117 "Ballard receives follow-on orders from Wrightbus for fuel cell modules to power 50 buses in UK", *Green Car Congress*, 10 March 2020. Available at: www.greencarcongress.com/2021/03/20210310-ballard.html.

118 "Emissions-cutting trucks and next-gen hydrogen buses closer to hitting the road with £54 million government-led funding", BEIS press release, 22 March 2021. Available at: www.gov.uk/government/news/emissions-cutting-trucks-and-next-gen-hydrogen-buses-closer-to-hitting-the-road-with-54-million-government-led-funding.

119 *Ibid*.

120 See: CMS, *Hydrogen law and regulation in Japan*, CMS Expert Guides 2021, available at: https://cms.law/en/int/expert-guides/cms-expert-guide-to-hydrogen/japan; and "Japan keeps auto industry's hydrogen dreams alive", S&P Global Market Intelligence, available at: www.spglobal.com/marketintelligence/en/news-insights/latest-news-headlines/japan-keeps-auto-industry-s-hydrogen-dreams-alive-62160857.

121 See: "Hydrogen Market in Japan", white paper for Connecting Green Hydrogen Japan 2021 event (12–13 October 2021). Available from: www.greenhydrogenevents.com/cghj.

122 See: "Japan: Strategic Hydrogen Roadmap", New Zealand Foreign Affairs and Trade, 30 October 2020. Available at: www.mfat.govt.nz/br/trade/mfat-market-reports/market-reports-asia/japan-strategic-hydrogen-roadmap-30-october-2020/.

123 "Cool Earth – Innovative Energy Technology Program", METI, March 2008. Available at: https://policy.asiapacificenergy.org/sites/default/files/CoolEarth.pdf.

124 Ministry of Economy, Trade and Industry (METI), Japan, The Strategic Road Map for Hydrogen and Fuel Cells. Available at: www.meti.go.jp/english/press/2019/pdf/0312_002a.pdf.

125 See: "Hydrogen Market in Japan", white paper, *supra*.

126 METI, The Strategic Road Map for Hydrogen and Fuel Cells, *supra*.

127 "Honda and Isuzu agree to conduct joint research on fuel cell trucks", *The Japan Times*, 16 January 2020. Available at: www.japantimes.co.jp/news/2020/01/16/business/corporate-business/honda-isuzu-agree-conduct-joint-research-fuel-cell-trucks/.

128 See: Toyota Motor Corporation, "JR East, Hitachi and Toyota to Develop Hybrid (Fuel Cell) Railway Vehicles Powered by Hydrogen", 6 October 2020. Available at: https://global.toyota/en/newsroom/corporate/33954855.html.

129 Kyle Martin and Ed Smith, "The energy investment landscape: the role of hydrogen in a decarbonised energy system", Lane Clark & Peacock LLP (LCP), 1 April 2021. Available at: www.lcp.uk.com/our-viewpoint/2021/04/the-energy-investment-landscape-the-role-of-hydrogen-in-a-decarbonised-energy-system/.

130 See: "Air Liquide inaugurates HyBalance pilot site in Denmark for production of carbon-free hydrogen", Green Car Congress, 8 September 2018. Available at: www.greencarcongress.com/2018/09/20180908-hybalance.html.

131 Ruven Fleming, "Clean or renewable – hydrogen and power-to-gas in EU energy law", *Journal of Energy & Natural Resources Law*, vol 39 no 1, 2020.

132 See: Directive (EU) 2019/944 (recast) of 5 June 2019 on common rules for the internal market for electricity. Available at: https://eur-lex.europa.eu/legal-content/EN/TXT/?uri= CELEX%3A32019L0944.

133 Ruven Fleming, "Clean or renewable – hydrogen and power-to-gas in EU energy law", *supra*.

134 Barney Cotton, "Hyundai Motor Company and Ineos to Cooperate on Driving Hydrogen Economy Forward", *Business Leader*, 23 November 2020. Available at: www.businessleader.co.uk/hyundai-motor-company-and-ineos-to-cooperate-on-driving-hydrogen-economy-forward/103563/.

135 "NortH2 welcomes new international partners RWE and Equinor", Gasunie news, 7 December 2020. Available at: www.gasunie.nl/en/news/north2-welcomes-new-international-partners-rwe-and-equinor.

136 Henry Edwardes-Evans, "Interview: Ryse Hydrogen's Suffolk freeport hydrogen vision takes shape", S&P Global Platts, 3 March 2021. Available at: www.spglobal.com/platts/en/market-insights/latest-news/electric-power/030321-interview-ryse-hydrogens-suffolk-freeport-hydrogen-vision-takes-shape.

137 London Assembly, "Mayor launches England's first hydrogen double decker buses", 23 June 2021. Available at: www.london.gov.uk/press-releases/mayoral/englands-first-hydrogen-double-deckers-launched.

138 See: Ministry of Economy, Trade and Industry, "Basic Hydrogen Strategy (key points)". Available at: www.meti.go.jp/english/press/2017/pdf/1226_003a.pdf.

139 IEA, "Global average levelised cost of hydrogen production by energy source and technology, 2019 and 2050". Available at: www.iea.org/data-and-statistics/charts/global-average-levelised-cost-of-hydrogen-production-by-energy-source-and-technology-2019-and-2050.

140 Angeli Mehta, "UK project aims to cut the cost of producing clean, green hydrogen", *Chemistry World*, 27 May 2020. Available at: www.chemistryworld.com/news/uk-project-aims-to-cut-the-cost-of-producing-clean-green-hydrogen/4011788.article.

141 See: Barbara S de Marigny, "Taxing for takeoff: The hydrogen economy in the US", *Petroleum Economist*, 26 April 2021. Available at: https://pemedianetwork.com/petroleum-economist/articles/sponsored-content/2021/taxing-for-takeoff-the-hydrogen-economy-in-the-us.

142 See: The White House, "Fact sheet: The American Jobs Plan", 31 March 2021. Available at: www.whitehouse.gov/briefing-room/statements-releases/2021/03/31/fact-sheet-the-american-jobs-plan/.

143 Angeli Mehta, "UK project aims to cut the cost of producing clean, green hydrogen", *supra*.

144 European Commission, "A hydrogen strategy for a climate-neutral Europe", *supra*.

145 E Napoletano and Benjamin Curry, "Environmental, Social And Governance: What Is ESG Investing?", Forbes Advisor website, 1 March 2021. Available at: www.forbes.com/sites/georgkell/2018/07/11/the-remarkable-rise-of-esg/.

146 See: Inter-American Development Bank (IDB), "Making Global Progress in Responsible Investments", 19 July 2019. Available at: https://idbinvest.org/en/blog/financial-institutions/making-global-progress-responsible-investments.

147 Sophie Baker, "Global ESG-data driven assets hit $40.5 trillion", *Pensions & Investments*, 2 July 2020. Available at: www.pionline.com/esg/global-esg-data-driven-assets-hit-405-trillion.

148 Minh Dao, Mark Spicer and David Dietz, "The rise of responsible investments", KPMG website, 30 April 2019. Available at: https://home.kpmg/xx/en/home/insights/2019/03/the-rise-of-responsible-investment-fs.html.

149 See: Regulation (EU) 2019/2088 of 27 November 2019 on sustainability-related disclosures in the financial services sector. (Note: the regulation came into force in March 2021.) Available at: https://ec.europa.eu/info/business-economy-euro/banking-and-finance/sustainable-finance/sustainability-related-disclosure-financial-services-sector_en.

150 Periodic reporting requirements under the SFDR will apply from 1 January 2022.

151 See: Practical Law Financial Services, "Sustainable finance: SFDR: overview", Thomson Reuters Practical Law. Available at: https://uk.practicallaw.thomsonreuters.com/w-023-4185?originationContext=document&transitionType=DocumentItem&contextData=(sc.Default)&comp=pluk&firstPage=true#co_anchor_a911216.

152 See: Vivien Shiao, "OCBC, riding growth wave, targets S$25b sustainable finance portfolio by 2025", *Business Times*, 22 June 2020. Available at: www.businesstimes.com.sg/banking-finance/ocbc-riding-growth-wave-targets-s25b-sustainable-finance-portfolio-by-2025.

153 See: HM Treasury, "A Roadmap towards mandatory climate-related disclosures", November 2020. Available at: https://assets.publishing.service.gov.uk/government/uploads/system/uploads/attachment_data/file/933783/FINAL_TCFD_ROADMAP.pdf.

154 European Commission, "A hydrogen strategy for a climate-neutral Europe", *supra*.

155 European Agreement Concerning the International Carriage of Dangerous Goods by Road (ADR), *supra*.

156 EU taxonomy for sustainable activities, *supra*.

157 IEA, "The Future of Hydrogen: Seizing today's opportunities", *supra*.

About the author

Dalia Majumder-Russell
Partner, Cameron McKenna Nabarro Olswang LLP
Dalia.Majumder-Russell@cms-cmno.com

Dalia Majumder-Russell focuses on energy transition technologies and infrastructure projects as part of the Energy and Climate Change practice in the CMS London office.

Dalia specialises in complex process energy projects advising on renewable energy transactions, commercial agreements and the regulatory frameworks within which such projects operate. For over 10 years she has advised governments, lenders and sponsors in Europe, Africa and North America on legal issues arising across the project lifecycle.

Dalia is passionate about the decarbonisation agenda across the energy, heat and transport sectors. She leads CMS's carbon capture usage and storage (CCUS) and hydrogen initiatives, and champions the firm's thought leadership on the subject which includes publication of the multi-jurisdictional *CMS Expert Guide to Hydrogen Law and Regulation*.

Dalia regularly speaks at industry events on hydrogen, CCUS and related matters and is a member of a number of industry expert groups helping catalyse the development of new regulation, legal framework and contractual standards in the sector.

About Globe Law and Business

Globe Law and Business was established in 2005. From the very beginning, we set out to create law books which are sufficiently high level to be of real use to the experienced professional, yet still accessible and easy to navigate. Most of our authors are drawn from Magic Circle and other top commercial firms, both in the United Kingdom and internationally.

Our titles are carefully produced, with the utmost attention paid to editorial, design and production processes. We hope this results in high-quality publications that are easy to read, and a pleasure to own. Our titles are also available as ebooks, which are compatible with most desktop, laptop and tablet devices. In 2018 we expanded our portfolio to include journals and Special Reports, available both digitally and in hard copy format, and produced to the same high standards as our books.

In the spring of 2021, we were very pleased to announce the start of a new chapter for Globe Law and Business following the acquisition of law books under the imprint Ark Publishing. We are very much looking forward to working with our new Ark authors, many of whom are well-known to us, and to further developing the law firm management list, among other areas.

We'd very much like to hear from you with your thoughts and ideas for improving what we offer. Please do feel free to email me at sian@globelawandbusiness.com with your views.

Sian O'Neill
Managing director
Globe Law and Business
www.globelawandbusiness.com